THE NATIONAL COUNCIL OF TEACHERS OF MATHEMATICS

THE ELEVENTH YEARBOOK

THE PLACE OF MATHEMATICS IN MODERN EDUCATION

BUREAU OF PUBLICATIONS
Teachers College, Columbia University
NEW YORK CITY
1936

Correspondence relating to and orders for additional copies of the Eleventh Yearbook and earlier Yearbooks should be addressed to
BUREAU OF PUBLICATIONS, TEACHERS COLLEGE,
COLUMBIA UNIVERSITY, 525 WEST 120TH STREET, NEW YORK CITY

Printed in the United States of America

MANUFACTURED BY THE HADDON CRAFTSMEN, INC.

CAMDEN, N. J.

CONTENTS

EDITOR'S PREFACE

THIS is the eleventh of a series of Yearbooks which The National Council of Teachers of Mathematics began to publish in 1926. The titles of the preceding Yearbooks are as follows:

1. A Survey of Progress in the Past Twenty-five Years.
2. Curriculum Problems in Teaching Mathematics.
3. Selected Topics in the Teaching of Mathematics.
4. Significant Changes and Trends in the Teaching of Mathematics Throughout the World Since 1910.
5. The Teaching of Geometry.
6. Mathematics in Modern Life.
7. The Teaching of Algebra.
8. The Teaching of Mathematics in the Secondary School.
9. Relational and Functional Thinking in Mathematics.
10. The Teaching of Arithmetic.

Bound copies of all except the first two Yearbooks can be secured from the Bureau of Publications, Teachers College, Columbia University, New York, N. Y., for $1.75 each postpaid. The first Yearbook is now out of print and the second is obtainable only in paper covers ($1.25 postpaid). A complete set of Yearbooks (Numbers 2 to 11 inclusive) will be sent postpaid for $14.25.

The purpose of the Eleventh Yearbook is to present as fairly and as clearly as possible the place that mathematics should hold in modern education. The book is not intended to be a defense of the teaching of mathematics, for when properly taught mathematics needs no defense. However, there is so much loose thinking among educators and laymen generally with respect to the importance of mathematics in the schools that it has seemed wise to attempt to clarify some of the most important issues.

I wish to express my personal appreciation as well as that of The National Council of Teachers of Mathematics to all of the contributors to this volume who have given so freely of their time and interest in helping to make this Yearbook so worthwhile.

W. D. REEVE

THE NATIONAL COUNCIL OF TEACHERS OF MATHEMATICS

The National Council of Teachers of Mathematics is a national organization of mathematics teachers whose purpose is to:

1. Create and maintain interest in the teaching of mathematics.
2. Keep the values of mathematics before the educational world.
3. Help the inexperienced teacher to become a good teacher.
4. Help teachers in service to become better teachers.
5. Raise the general level of instruction in mathematics.

Anyone interested in mathematics is eligible to membership in the National Council upon payment of the annual dues of $2.00. The dues also include subscription to the official journal of the National Council (*The Mathematics Teacher*). All correspondence relating to editorial business matters should be addressed to *The Mathematics Teacher*, 525 West 120th Street, New York City.

THE PLACE OF MATHEMATICS IN MODERN EDUCATION

ATTACKS ON MATHEMATICS AND HOW TO MEET THEM

By W. D. REEVE
Teachers College, Columbia University

Criticisms of Mathematics

It is my purpose in this chapter first, to present a number of representative adverse criticisms of the teaching of mathematics with some evaluation of their fairness and worth; second, to offer some of the most important and typical testimonials and constructive criticisms on the value and method of mathematics instruction; third, to indicate what seems to me to be the main causes of recent attacks on mathematics; and finally, to suggest certain improvements that will meet these criticisms.

In a recent discussion between two prominent educators concerning shortcomings in current school practices, one of them remarked: "Now take mathematics, for example; the trouble is that it never gets back into life—even the arithmetic or the other subjects that we teach in the junior high school." Can we deny that, on the whole, the statement is true?

Professor David Snedden, a confirmed critic of the practices in our secondary schools, holds that "algebra taught in American high schools is a nonfunctional and therefore nearly valueless subject for 90 per cent of all boys and 99 per cent of all girls—and that no changes in method or in content will change the situation." To what extent is this charge warranted by the facts?

Algebra and geometry. An editorial writer in a metropolitan newspaper offers the following criticism of algebra:

> Those who have followed the sailing chart of this column know that its conductor never assails public or private schools, their teachers or curriculum. It has criticized teaching methods and learning processes and will continue to do so.
>
> This column has, however, one bone to pick with the schools: *algebra*. Quite frankly I see absolutely no use for algebra except for the few who will follow engineering and technical lines. My confi-

dence in those who determine the curriculum will never be complete as long as they leave this subject in nearly every course of study.

I cannot see that algebra contributes one iota to a young person's health or one grain of inspiration to his spirit. I can see no use for it in the home as an aid to a parent, a citizen, a producer, or a consumer.

If there is a heaven for school subjects, algebra will never go there. It is the one subject in the curriculum that has kept children from finishing high school, from developing their special interests and from enjoying much of their home study work. It has caused more family rows, more tears, more heartaches, and more sleepless nights than any other school subject.

Algebra is required in practically every course except those courses which are frankly dumb-bell courses. It is a requirement for graduation; it is a requirement for college entrance, and yet in the face of all this it is not a requirement in any respect for anything in life with the one exception noted above.

I would not mind if it weren't for the fact that behind the question of algebra there is the larger question of academic tradition which believes that certain traditional subjects, long known to be useless, are still to be retained if one is to be "educated."

Now cut this out and hand it to your algebra teacher. I want her (or him) to write me and I want you to write. I am stripped for the battle.

It is obvious that algebra did not get back into the life of this writer. Moreover, if we continue to teach algebra as we have taught it in the past, we may expect the subject to pass out of the picture in most secondary schools.

Heywood Broun, in his column in the *New York World-Telegram*, recently wrote:

There must be something wrong with the manner in which we teach the young idea to shoot when such a large number of pupils hate their studies with so palpable a passion. . . . I have a vague impression that some of the higher mathematicians have announced that the axiom is no longer true. Einstein and others have brought curves back into favor. And all this irritates me no end because, whether it is true or false, my difficulties in acquiring the hypothesis have borne no later dividends. I cannot think of a single tough spot in my existence in which Euclid reached down to lend me a helping hand.

Just so! And how many boys studying geometry today will later give the same kind of testimony? What is responsible for such antipathy against mathematics? I think it is largely due to the stupid way in which mathematics is too often presented to the pupils.

In the February, 1934, issue of *McCall's Magazine*, in an article on "The Little Red School House," Maxine Davis wrote:

> Experiments show that many children who are compelled to study algebra get little or no value from it. On the contrary, it arouses irritation and annoyance, and a desire to escape school. How often have we heard our parents say, "Your school days are the happiest days of your life!"
>
> Well, mine weren't. There was that eternal bugaboo of geometry and trigonometry.

How can we expect people who have had this sort of emotional experience with the study of mathematics to be anything but critical? Some may say such persons are exceptions, but I fear they are not. When I told a woman who sat near me at a recent dinner that I was a mathematics teacher, she said, "How in the world can you make that subject interesting to children?" I replied that the task was very simple, but I do not think she believed me.

Mathematics is even condemned in four recent plays in New York City. In one play the line, "Why, she is as crazy as my geometry teacher," probably conveys more than was intended. In another play the comment is more rancorous: "I haven't thought of suicide since I was fourteen and failed in algebra."

Pitkin says, in his book, *Life Begins at Forty*:

> Everything is done to prevent life after twenty. Algebra is taught under compulsion to millions who can never use it even in intellectual play.

Life needs. On several occasions we have been told by educators that 85 per cent of the arithmetic we teach to children could be eliminated without loss to those who have to study it; that all they really need to know is whatever arithmetic is necessary in buying goods, making change, reading, writing letters, and traveling. In commenting on such remarks the editor of the *New York Times* on May 2, 1932 said:

> It is probably true that arithmetic in the schools, as one critic asserts, is cluttered up with too many branches having no practical value in modern life and of very doubtful "disciplinary" value. But when he goes on to include among the deadwood of the arithmetic textbooks the subject of "arithmetical ratios," this particular critic is in error. What the world needs at all times, and particularly now, is a sense of proportion, a sense of ratio.
>
> About percentage, for instance, one finds astonishing ignorance in the most unexpected quarters. Learned articles on the economic situ-

ation will speak of a decline of 100 per cent in employment or in foreign trade, which literally would mean the complete disappearance of employment or trade. To double anything is to increase it 100 per cent. To cut anything in half, which is the obvious meaning in the present case, is to reduce it 50 per cent.

The mad days of three years ago and the sad days we are living through now are both in a large measure attributable to a state of mind which has parted company with every law of arithmetic to enter a curved-mirror land of distorted ratios and proportions.

Here, again, is evidence that mathematics does not get back into life as it should.

The following[1] is typical of the kind of criticisms that appear today in our current periodicals:

> This problem of weighing the values of the subjects of the school becomes acute during depression. With bewildered taxpayers demanding economy and elimination of the nonessentials all school material should be given a fair hearing with all the evidence considered. Our needs today are not those of yesterday.
>
> A change along this line will not require much expense. In good-sized schools there will be found many sections of such subjects as Latin, algebra, physics and French. And a searching examination would prove that not more than one in five or even ten in those classes is getting a worth-while return from study as it is taught at present. The remedy would be to scrap all but one or two divisions of those subjects and have both teachers and pupils devote their time to something in the nature of frills and fancies. Teach the girls how to choose what they must buy; teach them little matters about making a home attractive. Teach the boys how to take care of a flower garden or a lawn; teach them how to make things with their hands. In short, train them to do what thousands who have gone before them are actually doing, not what seemed good to our grandfathers to teach our fathers.

These writers fail to recognize that many of our most capable boys and girls are as little interested in the so-called "frills and fancies" as others are in Latin and algebra. They usually like mathematics when it is properly presented. In fact, the solution to the problem lies in better-trained teachers and not in the elimination of fundamental fields of knowledge.

In the *Educational Research Bulletin* of Ohio State University for November 15, 1933 (p. 348), Charters says:

> There are scores of boys and girls, compelled to stay in school by law, to whom work upon civic projects would be an ineffable release.

[1] Farr, Henry L., *School and Society*, Vol. 37, p. 148.

And even for the average student it can easily be argued that facility in handling civic affairs of interest is of much greater value than a few extra theorems in algebra or an additional theme in English.

Another type of criticism may be illustrated by the following statement by Morgan in the *Journal of the National Education Association* for May, 1932:

> Increasing leisure demands richer school program. The stability and well-being of American civilization demand that our increasing leisure be matched by training in the arts of life. This requires an enriched curriculum in the schools. The most vital phrases of the curriculum under the new conditions are literature, music, and art. These are the real fundamentals. They are not fads and frills. The fads and frills are Latin and highly technical little-used phases of mathematics, such as square root, cube root, and apothecaries' weight. If the development of material power is not matched by a cultivation of the aesthetic and spiritual phases of life, our civilization must eventually crash.

Apparently this critic does not know that the topics he is attacking have already disappeared from the most of our schools. The damage to mathematics, however, is appreciable because the unenlightened public that struggled with these topics believes we are still emphasizing such topics in our mathematics classes.

Instructional weaknesses. At the 1932 Conference on Secondary Education, Dr. Jesse Newlon made the following statement:

> The American secondary school is a stronghold of conservation, where the curriculum is hopelessly traditional and not vitally connected with the needs of youth. The inertia of the secondary school is tremendous. Its teachers and administrators, with few exceptions, are but stereotypes of an outworn concept of education. Every attempt to affect fundamental changes has been stubbornly resisted by a majority of its personnel.

This is a caustic criticism with a large measure of truth in it; and while it does not single out mathematics, we can be sure that mathematics teachers must answer the charge. The fact is that the senior high school is the most reactionary part of our whole educational system with the possible exception of the colleges.

According to President Butler of Columbia University:

> The ability to perform the simplest mathematical operations is, to all intents and purposes, confined to teachers of mathematics or specialists in that subject. Algebra and geometry, whether plane or social, are as unfamiliar as the Laws of Manu.

President Butler made just as critical comment about reading, writing, and natural science, but that is no reason why we should congratulate ourselves.

Failures in mathematics. According to a report of the superintendent of the New York City schools in 1929, more high school pupils failed in mathematics than in any other subject. In one particular school more than half of the pupils failed in first-term algebra, and failure in high school mathematics as a whole was 26.9 per cent, the next greatest failure being in foreign languages. The report failed to mention that algebra is required of all students and is taken in the ninth grade, whereas some of the other subjects like physics and history are electives and are taken later. Nevertheless, one is not justified in believing that in the long run such large percentages of failure can be productive of good.

Algebra, studied by 56.3 per cent of the high school students in 1900, dropped to 35.2 per cent in 1928. The corresponding figures for geometry were 27.4 per cent and 19.8 per cent. These decreases are doubtless greater today. While some of this loss can be attributed to the introduction of new subjects into the secondary schools, part of it is undoubtedly traceable to a genuine loss due to a dislike for the subject. The question naturally rises here as to whether the fact that a large number of children of eligible age are not now in high school may not be in itself an implied criticism of the subjects in the curriculum, algebra included.

In a recent report of an investigation of the mathematics necessary to do acceptable work in college courses in educational statistics in the United States, it was pointed out that simple arithmetical and algebraic tasks are not easy for many graduate students. A large percentage of almost 1,000 students who took an inventory test were utterly unable to do some very simple exercises involving the essential abilities for successful mastery of educational statistics. Many of these students, and students in other fields as well, have for some reason or other developed an unfortunate emotional complex against these simple elements of mathematics. I believe this is partly due to the non-mathematically minded teachers who should never have been permitted to teach the subject.

Curricular trends. Until recently all undergraduates at Princeton University were required to continue the study of either the classics or mathematics in their freshman year. Under the new regulations it is permissible to substitute other subjects for this requirement. Commencing this year, Barnard College is omitting intermediate algebra from its list of required entrance subjects. It is clear that changes in requirements will be made in other colleges and universities. Such changes are all a part of the present movement to lessen the emphasis that has been placed upon mathematics in the schools.

A Midwestern superintendent of schools, after attending a conference of superintendents in Chicago, made the following statement to his teachers of mathematics:

> Since only 12 per cent of the pupils who study mathematics in the high school are ever going to college, we do not need to worry any longer about teaching mathematics in the secondary school.

Such changes in administrative policies and the persistent criticism of mathematics challenge the very existence of mathematics in the curriculum of the secondary school. Some teachers of mathematics are inclined to take an indifferent attitude toward the question. They say "If anything is to be dropped let it be the fads and the frills." One teacher remarked, "We should worry, we are no worse than the teachers of English." Perhaps not, but the question cannot be dismissed in this way. My own feeling is that the present complex state of affairs in which we must continue to live demands a greater knowledge of mathematics rather than less. However, anyone who thinks that mathematics cannot be dropped from the high school curriculum needs only to recall the fate of formal grammar and Greek to realize what is possible. As one commentator puts it, "If a pupil postpones his Greek until he enters college today, why should he not postpone his Latin, his algebra, and his geometry tomorrow?"

Nor can we dismiss the attacks by ridiculing the critics. They may be and often are ignorant of mathematics and in some cases have been the victims of non-mathematically minded teachers; but they are not stupid. They know that the teaching of mathematics is not what it should be and they are determined either to bring about an improvement or to drive mathematics entirely out of the curriculum.

Who shall study mathematics? We are all aware that the pupils in our secondary schools are not equally gifted in mathematical abilities. Some individuals should certainly be excused from further study of the subject, but no one, I think, has the wisdom to decide who will profit most by its study or to predict who the future Newtons and Einsteins are to be. David Eugene Smith's delightful stories of the unpromising youths who later became prominent mathematicians point the moral in this instance.[2] I quote:

. . . Of late a new type of educator has appeared, the one who proposes to weigh in psychological scale the intellect of youth and to guide it aright. . . . [We] have it today in the phase of vocational guidance, and in this work so many excellent people are seriously engaged that we are certain to see it become an important phase of modern education. Let the boy who gives promise in science begin his specialization early, say those who seek to guide the youth in a scientific manner, and let the one who takes to Latin bend his energies there. Let there be scientific tests to show whether or not the particular individual can hope for success in the particular vocation—a worthy effort and one that will produce good results. But there are not wanting those who will be less scientific, and who will assert that one who, by virtue of his surroundings and family, is destined to be a hewer of wood, should early come to like to hew, and should be taught chiefly the nobility of labor with the hand. That we may realize some of the dangers that beset those who seek to guide the youth aright, and who may feel called upon to sidetrack all that is not immediately practical, let me tell you some advice that I myself have given within a few years past in cases like these, and lay before you the problem that I had to face.

Not long ago there came to me a father who wished to train his boy for trade in a seaport town, and who asked my advice as to the proper education to give him. The problem seemed simple. The community was not an educated one; it lived off its little shipping industry; the boy was destined to small business and to small reward; he gave no promise of anything better, and the advice was, therefore, unhesitatingly offered that the only mathematics he needed was arithmetic through the sixth grade.

Another parent asked me a little later about his son. The boy was of the ordinary type and would probably follow his father's occupation, that of a sculptor. What mathematics would it be well for him to take? I suggested a little study of curves, some geometric drawing, and the modeling of the common solids—a bit of vocational guidance that seemed to me then and seems even yet particularly happy.

A third boy happened to be with me on a steamer and I took some interest in talking with him and with his mother. They lived in a city of no particular note, at any rate at that time, and the boy was

[2] Smith, David Eugene, *Problems in the Teaching of Secondary Mathematics*. An address delivered before the New England Association of Teachers of Mathematics. Ginn and Company, 1913.

going into the selling of oil within a few years. The profits of the Standard Oil Company appealed to the family, and I advised him to learn his arithmetic well and get into business as soon as he could.

Out of the store of my memory I recall a curious lad whom I came to know through my sympathy with the family. The mother was a poor woman, and she took the boy, when little more than a baby, over to Riverside Park one day when there was a naval parade. A drunken sailor, having had a fight with a group of hoodlums, rushed through the crowd of spectators and slashed right and left with a knife. In the excitement the boy, in his mother's arms, was horribly cut in the face. When I knew them he was about ten years old, unable to speak plainly, and already a misanthrope through his affliction. I advised the mother to give the boy a vocational education, telling her that through the use of the hands he would satisfy his desire for motor activity, and that this would compensate him for the loss of verbal fluency and would tend to make him more contented with his lot. In this advice I feel that I would have the approval of educational circles.

And finally, out of this series of experiences, let me recall the case of a boy whom I came to know through a noble priest who found him one morning, an infant a few days old, on the steps of his church. We talked over the best thing to do for such a foundling, one who, at the time I knew him, was in the primary grades. He showed no great promise, he was without family recognition, and his only chance, apparently, was in the humbler walks of life. I recommended a vocational school where he could quickly prepare for the shop or the lower positions of trade, and the good priest approved my plan at the time, although he finally followed quite a different course.

It is apparent, however, that I have here spoken in parables. Perhaps you already recognize the boys, and perhaps you feel how sadly I blundered in my counsel. For the first of these whose cases I have set before you felt a surging of the soul a little later, and this was recognized in time, and he became one of the Seven Wise Men of Greece—Thales the philosopher, he who introduced the scientific study of geometry into Greece. The second felt a similar struggle of the soul, and his parents recognized my poor counsel in time to save him and to give to the world the founder of its first university—Pythagoras of Samos. The third boy, for whom only the path of commerce seemed open, and this in a town only just beginning to be known, was the man who finally set the world's first college-entrance examination, the one who wrote over the portal of the grove of Academos the words, "Let no one ignorant of geometry enter here"—Plato, the greatest thinker of all antiquity. The fourth, the hopeless son of poverty, maimed, sickly, with no chance beyond that of laboring in the shop for such wage as might by good fortune fall to his lot, became the greatest mathematician of his day—always the stammerer (Tartaglia), but one whom Italy has delighted to honor for more than three centuries. And the last one of the list, the poor foundling on the steps of St. Jean-le-Rond in Paris, became D'Alembert, one of the greatest mathematicians that France, a mother of mathematicians, ever produced.

Shall we, then, advocate the selection of those who are to study mathematics and close the door to all the rest? Are we so wise that we can foresee the one who is to like the subject, or succeed in it? Have we so adjusted the scales of psychology that we can weigh the creases in the brain, or is there yet invented an X ray that will reveal to us the fashioning of the cells that make up its convolutions?

Of course it will at once be said that these illustrations that I have given are interesting, but that they are unfairly selected; that those boys gave earlier promise in mathematics than I have said. It will be asserted that I should have taken the case of the stupid boy, the one who did not like school, the one who liked to play with little wind wheels, who liked to fight, who actually did run away from school, and who stood near the bottom of the class in mathematics. Such a case would be a fair one, one in which we could safely say that prescribed algebra and geometry are out of place. And I suppose we must agree to this and confess that the argument from the historical incidents that I have mentioned was unsound. Let us rather take this case that I have just described, and let us see to whom the description applies. I need hardly tell you who this boy is; he is well known to you; he is well known to the world; and long after every educational reformer has passed into oblivion his name will stand forth as one of England's greatest treasures, for it is the name of Sir Isaac Newton.

But again I have been unfair, perhaps. I should have taken positively hopeless cases, for such can surely be found. I should have taken some illiterate man, one who does not learn to read until he is nearly out of his teens, or else some man who shows no promise in mathematics by the time he reaches manhood, or some one who by the time he is thirty is to show no aptitude in the science. It is so easy to theorize! But let us have care, for the men whom I have now described are Eisenstein, Boole, and Fermat. Take them away and where is your theory of invariants, your modern logic of mathematics, and the greatest genius in theory of numbers that the world has ever seen?

Indictment against secondary education. A recent criticism made by Professor Thomas H. Briggs[3] against secondary education may, I think, be particularly applicable to mathematics.

Partly as a result of the impossibility of making the curricular adjustments that are obviously needed, partly because of recurrent and resounding criticisms from those who find it easier to be destructive than to build up a new program, many teachers have lost faith in the efficacy of the very subjects that they profess to teach. This is one

[3] Briggs, Thomas H., "A Vision of Secondary Education," *Teachers College Record,* October, 1932, pp. 7-9.

of the most tragic and discouraging facts that affect secondary education. How can a person teach with enthusiasm and effectiveness that in the efficacy of which he does not have complete faith? A lack of confidence results in small and decreasing knowledge, in perfunctory teaching and in meaningless drudgery that controverts values.

For this and for several other reasons the students in our secondary schools understand, learn, retain, and use little of what the curriculum plans. The facts are abundantly presented in our professional literature and may be observed by anyone who takes the trouble informally to test a fair sampling of the schools. By any reasonable criterion the investment by the public in our secondary education has failed to pay academic and social dividends that justify the huge expenditures of time, effort, and money. By any reasonable audit secondary education for the masses is bankrupt. Inappropriate offerings make inevitable small accomplishment by those naturally incompetent for the traditional academic achievement, and the admission, encouragement, or compulsion of such pupils to pursue these subjects interferes with the progress of those who by natural and environmental gifts might be expected to attain significant accomplishments. Students most academically gifted are now most severely handicapped. Others in a hopeless struggle with what is inappropriate to their talents are achieving little that contributes to their own advancement or to that of society. It is difficult to emphasize this fact for the great majority of our youth and at the same time to recognize the eminent success for the small minority who in the best schools find the opportunity for achieving the peculiar success that should be common to all. In the success of this minority is the beacon that should lead the profession to make secondary education the most potent factor in preserving and advancing our democracy. The small, the pitiably small, accomplishment of the majority of our pupils is a well-established fact. Though perhaps an inevitable concomitant of an attempt rapidly to provide a secondary education for all, it is discouraging; but even more discouraging is the complacency with which the repeatedly published fact has been received. Perhaps this complacency exists only because we do not know what to do. Perhaps we merely await leadership to advance out of the wilderness.

With the rapid development of the number and enrollment of secondary schools we have not been able to recruit and adequately train a body of competent teachers. It would have been difficult under any circumstances to prepare in every decade almost as many new teachers as there were experienced teachers in service, but when the complexity of the student body steadily increased and the purposes of secondary education steadily became more confused, it was impossible. Attributing high praise to the top fraction, perhaps one-tenth, and generous credit to perhaps five-tenths more, the brutal fact is that approximately half our high school teachers are below, many of them far below, the standards that should reasonably be required. With all teachers in the French *lycées* and in the German *Gymnasiums* our small fraction of best can no more than hold their own. While training in methods of teaching and in the other appurtenant subjects of education is important, it can never take the place of thorough knowledge of the subjects to be taught. It should supplement rather than supplant. Our teachers are relatively strong in method, but limited and weak in knowledge. Too few of them in the academic fields exemplify the culture that liberal education should develop; too many are satisfied with the dangerously little learning that they have acquired in college courses, without at the same time acquiring an insatiable appetite that leads them ever onward in learning and in the enjoyment of learning. The public is largely to blame, for it has judged personality more than scholarship and effective inculcation of culture. Whatever the large program for the future of secondary education, it should begin in our colleges for teachers with a better training, adequately founded on knowledge and a love of knowledge.

Finally, there is everywhere, among the profession no less than among the public, an expectation of miracles. Although the fetish of education, however it be defined and however small its accomplishments in making its products more socially and intellectually effective, is dominant, there is a widespread suspicion that all is not as it should be in our secondary schools. Evidence of defects is hospitably heard, but little is done. There is generally a feeling that some panacea will be found, some single act that will remedy all defects and set us on the highroad to educational prosperity. This sentiment, it may be noted, is not peculiar to the field of education. Needless to say, there is no such panacea, nor will there ever be, in education or in economics. The causes of depression are complex; the cure must concern the fundamental. So long as we look for a miracle that will overnight give us a full-fledged and permanent program for effectiveness, just so long we shall endure inefficiency, squander vast sums of money and infinite possibilities of youth, and fall back in vain satisfaction with the fetish. We need a vision that will disturb our complacency, stimulate us to action, and direct the unending efforts toward educational betterment.

Reform in education. The late Dr. Henry Suzzallo in speaking of a need for a sweeping reform in public school education said: "There are too many academic specialists in our public schools who know very much about a little field of study and nothing much about the wider ranges which make up the rest of life." He expressed regret that such specialists in subject matter have been given preferment in appointment, promotion, and salary, with the result that departmentalized teaching has increased to the detriment of the child. The teacher of the future, said Dr. Suzzallo, should be "a character man, or woman, interested in people as well as in books, a man of this world and not an academic recluse."

I believe we could all agree with this point of view. How to remedy such a situation is a problem of great importance to us all. One method of handling this problem is to set up what is called an *integrated program.* I have discussed this rather fully elsewhere.[4]

Testimonials Favorable to Mathematics

Mathematics in government. Mark Sullivan, as a critic of the New Deal, laments that men able to use calculus and the higher differential equations are being brought to Washington for utilization of that ability in government service. He has attacked Ezekiel for attempting to state statistically the demand, supply, and price of hogs in the United States for some years ahead. In defense of the mathematician, however, Dr. Harry Elmer Barnes recently expressed the following opinion in the *New York World-Telegram.*

> It is possible to put a very different interpretation on these developments in Washington. To many it means the first real effort to bring government thoroughly up to date. The fact that men like President Cleveland had not the slightest grip on the mathematical and scientific facts which have created the modern world is one reason why we are in our present mess. Many other Presidents were even worse off in this regard. Even Woodrow Wilson, a university president, was blissfully innocent of modern science and engineering and relied upon the rhetorical methods of the age of Cicero.
>
> Our modern material civilization has been built up by the use of calculus and higher mathematics. It is not unreasonable to assert that a comparable intellectual equipment and apparatus are needed to control our institutional life. Without the calculus and higher differential equations we could not build a complicated machine, a skyscraper, a bridge, a subway, a railroad, a large ocean-going ship, an electric lighting plant, a telephone, a telegraph cable, radio or any other representative manifestation of modern ingenuity.
>
> If this be so, how can one expect to run a government, which is far more complicated than any machine or engineering project, without at least some appreciation of the function of higher mathematics? We may be glad that there is a man in the White House whose perspective has advanced beyond long division.
>
> It is high time that our statesmen should begin to think in terms of modern science instead of relying upon the rhetoric and hot air which have dominated politics from Pericles to many of the present spellbinders in the United States Senate.

Social values. The social values of mathematics, both realized and

[4] Reeve, W. D., "Mathematics and the Integrated Program in American Secondary Schools." *Teachers College Record,* April, 1935, pp. 497-507.

potential, are stressed by the mathematician, E. T. Bell,[5] in the following statement:

For it must now be obvious, even to a blind imbecile, that American mathematics and mathematicians are beginning to get their due share of those withering criticisms, motivated by a drastic revaluation of all our ideals and institutions, from the pursuit of truth for truth's sake to democratic government, which are only the first, mild zephyrs of the storm that is about to overwhelm us all. In the coming tempest only those things will be left standing that have something of demonstrable social importance to stand upon. Mathematics, we as mathematicians believe, has so much enduring worth to offer humanity on all sides from the severely practical to the ethereally cultural or spiritual, that we feel secure—until we stop to think.

The arresting thought that we as mathematicians have done next to nothing to inform and convince the sweating men and the sweated women, whose hard labor makes possible our own leisurely pursuit of "the science divine," that mathematics does mean something in their lives and might mean more, may well make us apprehensive of the future, for these too patient men and women in the storm ahead of us all will cast the deciding vote.

The harsh attrition has already begun. Are not mathematicians and teachers of mathematics in liberal America today facing the bitterest struggle for their continued existence in the history of our Republic? American mathematics is exactly where, by common social justice, it should be—in harnessed retreat, fighting a desperate rear guard action to ward off annihilation. Until something more substantial than has yet been exhibited, both practical and spiritual, is shown the non-mathematical public as a justification for its continued support of mathematics and mathematicians, both the subject and its cultivators will have only themselves to thank if our immediate successors exterminate both.

Taking a realistic view of the facts, anyone but an indurated bigot must admit that mathematics has not yet made out a compelling case for democratic support, so that the men and women who pay the bills which make mathematics possible can see clearly what they are asked to pay for. This must be done, and immediately, if mathematics is to survive in America.

Our task is to develop somehow a generation of teachers who not only know a great deal about their own field of study, but who also understand something of the "wider ranges which make up the rest of life." The trouble is not so much with mathematics as some of our critics would have us believe. I have elsewhere re-

[5] Bell, E. T. In a review of *The Poetry of Mathematics and Other Essays*, by David Eugene Smith. *The American Mathematical Monthly,* November, 1935, p. 559.

ferred to several authorities who have exalted mathematics, and one yearbook in the series of which the present volume is a part was devoted entirely to the topic "Mathematics in Modern Life."[6]

The *Columbus* (Ohio) *Dispatch*, commenting on the proposed curtailment of arithmetic in school curricula, maintains that the schools should not withhold the "mathematical key" to the "gateways of a larger life":

> One of the "educational experts" of Teachers College, in Columbia University, says of the study of arithmetic: "Arithmetic should be taught, but not all arithmetic. We should transmit to the children of tomorrow only that which they are likely to need. Even highly cultured people have need for only a small amount of mathematics."
>
> His conclusion is that something like 85 per cent of the arithmetic usually taught in the schools should be dropped out. He can see little use for anything more than simple addition, subtraction, multiplication, and division. A questionnaire sent to some 600 persons found but few to reply that they had found the higher mathematics of any direct benefit to them.
>
> This was quite natural, as a result of the questionnaire method. One would like to see the results of a questionnaire sent to 100,000 or more men and women who have been "questionnaired" for the past 20 years asking the one simple question, "Do you favor the questionnaire, as a means of getting comprehensive, accurate and reliable information on debatable matters of opinion or sentiment?"
>
> The difficulty with this Columbia expert's opinion as to the teaching of mathematics is that its practical application in the schools would leave a countless number of young people handicapped, later on, when they desired to push into one or another interesting and profitable life work, or subject of study, and found that they did not possess the mathematical key necessary to unlock the door. It is the place of education to open up, not close, the gateways of a larger life.

Commonplace values. In the opinion of Myron C. Taylor, prominent business man, mathematics is the most important subject in the curriculum:

> What shall we put into the mind? For the uses of the present day we could hardly put too much mathematics into it. A brain that is mathematically minded has a decided advantage over any other in affairs of trade and commerce; it is a fine type of mind to approach problems in physics, and it is just the sort of mind that would lead one to be philosophical. And so perhaps if one were permitted to study only one subject, depending upon companionship and one's everyday contacts with the world to pick up the language of the community after a fashion, mathematics would serve one better than any

[6] *Sixth Yearbook of the National Council of Mathematics*, 1931.

other subject in the courses of study. So it would seem that one cannot give too much attention to mathematics.

A recent trend in high school commencement programs is the vitalized commencement, in which the activities develop from the regular classroom work. The commencement program for Hiram High School (Hiram, Ohio) in 1929 "centered around the idea of mathematics in everyday life." A report of this program, by Mr. Harold E. Davis, appeared in the April, 1930 issue of *Ohio Schools.* The purpose of the program was:

> . . . to show the graduating class in the performance of work characteristic of the school and to establish some definite relationship between school work and the lives and activities of the students. It has been the purpose furthermore to *intellectualize* the commencement program, and to give the graduates a chance to show they have developed sufficient intellectual interests and capacity to talk interestingly to their parents and friends about some topic of wide interest and concern, to which they have given particular attention.

Edward Howard Griggs,[7] in his new book, *The Story of an Itinerant Teacher,* mentions the personal benefits that he has derived from the study of mathematics:

> Feeling that with my wide reading, I needed intellectual discipline, I chose mathematics for my major subject, with no intention of using it afterward, in teaching or other practical ways. Never have I regretted the choice. Mathematics is one of the oldest subjects in the curriculum, and one concerning which there is at present the most confusion of thought. It is the one completely exact science, just because it deals with pure ideal concepts. We define a straight line as the shortest distance between two points. There is no straight line in Nature. Nobody ever saw or drew a straight line. What we call straight lines are poor, approximate illustrations of an idea that exists only in the mind. Mathematics is thus pure reasoning with ideal concepts. When you argue that two apples and two apples make four apples, that is not mathematics; it is a concrete illustration. It is mathematics only when you think that two and two make four, to any intelligence, anywhere in the universe, any time throughout eternity. Mathematics is thus the grammar of science, a language for the statement of all scientific facts and laws. Anything that can be stated at all can finally be stated in mathematical terms; and only then is the statement exact.
>
> In our effort to introduce mathematics early in the course of study, we have brought in all sorts of concrete material, and permitted the pupils to memorize. Learning the multiplication is, of course, sheer

[7] *The Story of an Itinerant Teacher,* pp. 41-43. Copyright, 1934. Quotation used by special permission of the publishers, The Bobbs-Merrill Company.

memory work, and should be accomplished in the memory period, until one can say it forward and backward, awake or asleep. Everything else in mathematics is pure reasoning; and allowing the pupil to memorize problem and proposition, brings him, at eleven or twelve, up against a dead wall; and he finds that he does not like mathematics. Nine times out of ten when a student does not like mathematics, it means that he has been wrongly taught.

I am not one of those modern educators who hold that discipline in one field can not be carried over into other fields. Of course I know what they mean when they say that "knowing algebra and geometry will not help you to decide whether the butcher has cheated you on six pounds of steak"; but accurate thinking is accurate thinking; and mathematics is the one perfectly accurate form of thinking known to man. Once the habit of accurate thinking is formed, it applies to everything you think about. It is not necessary to speak theoretically; I know that I can organize my ideas more logically, better avoid overstatement and express my thought more accurately, because I did some six years of mathematics during my brief term in college.

We need more testimony like this to put before the people who are not convinced that the study of mathematics has any importance in the schools. But better still we need to teach mathematics so that our pupils will ultimately offer such testimony.

Mathematics and the cosmos. The following editorial statement appeared in the *New York Sun* on November 8, 1930:

In search for the ultimate reality in nature, now virtually abandoned as beyond the powers of the puny three-dimensional intellect of man, mathematics was long a useful guide. Its importance increased with time until men of science came to regard it as the only language in which a natural phenomenon could properly be discussed. Today the curious are told that they will know all that can be known about such a phenomenon when they have discovered a mathematical formula which adequately describes it. Mathematics is no longer content to be a tool; it threatens to absorb almost everything else. Sir James Jeans asks us to look upon the Great Architect of the Universe as a pure mathematician.

In the Rede Memorial Lecture at the University of Cambridge the other night the brilliant British astronomer expounded his latest physical and cosmic theories. He had expanded the lecture into a small volume, *The Mysterious Universe,* which he put on sale in the book shops the next day. Like everything which comes from his pen, it has a fascinating boldness. There is danger of some of his readers taking his ideas for scientific gospel. But he warns them. "Everything that has been said," he writes in his final chapter, "and every conclusion that has been tentatively put forward, is quite frankly speculative and uncertain." The conclusions, except in the last chapter, are not wholly or principally his own; he reviews the whole field of modern physical science.

The universe begins to look to Jeans more like a great thought than a great machine. If it is a universe of thought, then its creation must have been an act of thought. "Time and space," he says, "which form the setting for the thought, must have come into being as part of this act." The line of demarcation between realism and idealism is becoming blurred. Objective realities exist, but we are assuming too much if we label them either "Real" or "Ideal." The true label he thinks is "mathematical," but this term must be taken to connote the whole of pure thought and not merely the studies of professional mathematicians.

What does the man in the street gather from it all? Perhaps that science shows signs of becoming religious. Pure materialism is rather out of date. When matter is dissected it is found to be composed of something which does not meet our ideas of matter at all. Nobody nowadays attempts to say just what an electron is or ventures to hope that man can ever know what it is. As baffling as the mysteries of the microcosm are those of the macrocosm. Suns and planets may evolve out of nebulae, and it is not difficult to picture the birth of nebulae from the great agglomerations of atoms; but whence the protons and electrons which compose them? The mathematician may put these questions aside as idle and impractical and find consolation in his beautiful equations; the rest of the world will not so easily be satisfied.

Functional thinking. The following excerpts are from a recent study by Dr. H. R. Hamley, in which he discusses mathematics as of mode of thinking.[8]

Our purpose in quoting these authorities [Spearman, Dewey, Rignano and Piaget] is partly to uphold the thesis that reasoning is ideal experiment, and partly to support the view that school mathematics should be taught as the symbolic expression of actual or potential activity; in other words, that school mathematics should be presented as a concrete and dynamic, rather than as an abstract and static, science. We do not think that the function concept can be grasped by the average student in any other way. Mathematics is the projection of *life* upon the plane of human imagination.

Now, the conduct of experiment involves the following main activities: the collecting of data, the arrangement of the data according to some attribute or quality, the identification of the data by name, and the interpretation of the data so organized in terms of relations or correlations that have been found to subsist between them. In other words, experiment involves the recognition of the mathematical concepts: *class, order, variable,* and *correspondence.* . . .

[8] "Functional Thinking," *Ninth Yearbook of the National Council of Teachers of Mathematics*, pp. 37-38, 44-46.

. . . We have now to inquire how far these aims are capable of realization through functional mathematics. Of the utilitarian value of functional mathematics there can be no question. The function is the mathematical correlate of physical change, expressing in symbolic language the relationships that accompany change in the physical world about us. . . . the utilitarian value of functional mathematics is commensurable with the utilitarian value of physical progress.

The cultural value of mathematics can hardly be considered apart from that of mathematics in general. D. E. Smith has appraised the value of mathematics in an article both eloquent and profound. He shows that mathematics did not come into being to satisfy utilitarian needs. "It seems rather to have had its genesis as a science in the minds of those who followed the courses of the stars, to have had its early applications in relation to religious formalism and to have had its first real development in the effort to grasp the Infinite." That mathematics has beauty has not been generally appreciated. Few who are not devotees would be ready to claim that "the true spirit of delight, the exaltation, the sense of being more than man, which is the touchstone of the highest excellence, is to be found in mathematics as surely as in poetry. . . ." If beauty is an attribute of culture, then cultural mathematics can claim to provide, in a special degree, the food of culture, for order, rhythm, symmetry, harmony and unity, which are among the accepted qualities of beauty are also among the popular concepts of functional mathematics.

That mathematics has disciplinary value some are disposed to doubt. They maintain that the days when it was believed that accuracy, judgment and reasoning were specific abilities that could be developed by appropriate training and discipline have long since passed away. Others, while admitting that the doctrine of formal discipline has been rudely shaken, still cling to the belief that there is something in "the human worth of rigorous thinking." This subject was considered to be a matter of such vital importance that the National Committee on Mathematical Requirements made a special study of "The Present Status of Disciplinary Values in Education," giving the result of the experiment and inquiry up to the year 1922. Many experimental studies of this subject have been made since 1901, when Thorndike and Woodworth undermined our faith in the faculty doctrine. Some of these have been vitiated by faulty technique, others have been discounted because they do not touch the real problem. Those that may be considered valuable and unobjectionable as to procedure seem to point to a conclusion which may be summed up in the words of Burt: "A common element is more likely to be usable if the learner becomes clearly conscious of its nature and of its general applicability: active or deliberate transfer is far more effective and frequent than passive, automatic, or unintentional transfer. This seems especially true where the common element is an element of method rather than of material, an ideal rather than a piece of information." In other words: method of procedure, consciously accepted as a desirable ideal, is the key to the problem of transfer.

The bearing of this conclusion on our present problem will be obvious, when we remember that functional mathematics is neither a technical skill nor a formal discipline but a *mode of thinking*. It may, and often will, involve many skills but its real domain is to be found in the concomitant changes of correlative variables and the relations that subsist between them. If we seek the material of transfer in the common elements of mathematics and life, we find them in the concepts: class, order, variable, relation, correspondence, correlation, and function.

Numerous reasons might be given to explain why mathematics has come in for so much criticism in recent years, particularly algebra and geometry in the secondary school, but I will mention three which I consider of fundamental importance. In the first place, the stupid way in which many teachers continue to present both algebra and geometry contributes largely to the unpopularity of these subjects. So long as we continue to teach these subjects mechanically and include in the course of study much material that is obsolete, we may expect the pupils to dislike mathematics.

Secondly, the general educators, many of whom perhaps have had unfortunate experiences with mathematics in school, consider that it is the subject *per se* that is at fault. They would therefore eliminate it. There are some who, although they may not know much about the subject, or even how it should be taught, recognize its intrinsic worth while at the same time they know that the teaching of mathematics is bad in many schools.

In the third place, the professors of pure mathematics in the colleges and the universities who once had whatever responsibility there was for training their students to be teachers of mathematics turned over the responsibility for such work to the schools of education. The results reflect the obvious lack of interest. Schools of education and teachers colleges have dominated the teacher-training programs, and instead of turning out teachers with proper academic background they have produced too many teachers who know but little more subject matter, sometimes less, than what they were actually supposed to teach. Standards for teachers have thus become topheavy with requirements in general methods courses, principles of teaching, and the like, and with practically no uniform academic standards that are satisfactory.

The problem of mathematics education in this country cannot be solved until we remedy at least the three situations first described. We must first set up higher standards for teachers of mathematics.

For example, no teacher should be allowed to teach mathematics, even in the junior high school, who has not at least completed a substantial course in the calculus.

In the second place, we must educate the educators with reference to the importance of their supporting our higher standards for teachers and also with respect to the importance of mathematical education in the schools. This can be done if we show them that the modern course in mathematics is modeled on the 1936 plan instead of that of the seventeenth century.

Finally, we must convince the professors of mathematics in the colleges and universities that they must take a more prominent part in the teacher-training program. Most of their students will probably plan to teach mathematics. It is the duty of such college and university teachers to pay more attention to the future needs of their students and help the rest of us, whose primary function is to train our students to be teachers.

THE REORGANIZATION OF SECONDARY EDUCATION*

A STUDY OF PRESENT PROBLEMS AND TRENDS, WITH PARTICULAR REFERENCE TO MATHEMATICS

By WILLIAM BETZ
Rochester, New York

"The day has come to begin the work of our renovation . . . We must arise and move on.. We must liberate ourselves from blind technology and grasp the complexity and the wealth of our own nature."—Alexis Carrel

Introductory Statement

Facing a staggering problem. Never in the whole history of education have teachers been more baffled and worried by the scene they are obliged to face from day to day. So great is the confusion of the present period, and so vexing are the questions demanding settlement in every field of human thought and endeavor, that one shrinks with some dismay from a survey of the educational programs set up for our adolescent youth. Yet, so pressing is the need of orientation, so vital is an honest formulation of our difficulties, as well as a re-examination of our basic faith in a system of universal education, that every effort in the direction of a comprehensive review, however fragmentary and halting, is at least better than passive or stagnating indifference. In the following pages the attempt will be made to lift into prominence and to analyze those factors which seem to be mainly responsible for our present educational troubles. With the analysis will be submitted constructive criticisms and tentative proposals on behalf of a desirable co-ordination of the old and the new. In order to remove the appearance of a purely subjective appraisal, there will be frequent references

* The writer wishes to express his indebtedness to Professor W. D. Reeve of Teachers College, Columbia University, for valuable editorial assistance, and to the numerous publishers and authors who made possible the quotations and references appearing in this monograph.

to the current educational literature, thus enabling the reader to check the author's position against a wider body of opinions.

Nature of questions to be discussed. The extent of the territory to be covered is such that it will be impossible to include in this survey anything but the "high spots" of the educational situation. The following queries may serve to suggest the general nature and scope of the problems to be discussed:

1. What are the chief causes of the present educational unrest?
2. Have we a consistent and practicable philosophy of education?
3. To what extent should "social reconstruction" be a determining factor in a basic philosophy of education?
4. Which guiding principles, if any, may be applied in the appraisal of the newer educational trends?
5. What is the present status of the curriculum revision movement?
6. How are the secondary schools trying to cope with the problem of mass education?
7. Should vocational education be encouraged at the secondary level?
8. Is it possible to maintain standards if secondary education is to be adapted to the individual pupil?
9. Along what lines should classroom procedures and methods be changed, if the findings of the newer psychology of learning are accepted?
10. Does the professional preparation of secondary teachers call for extensive readjustments?

Although this study is intended primarily for teachers of mathematics, its scope must necessarily be enlarged to include the educational scene as a whole. Education is viewed as a unified process extending from the kindergarten to the university. Hence, frequent excursions will be made into the elementary field for the sake of a more inclusive perspective.

PART ONE

EDUCATION IN A CHANGING WORLD

A World in Turmoil

External causes of world-wide unrest. That the whole civilized world is now going through an upheaval unparalleled in history has become a commonplace observation. From day to day the newspapers bring us startling accounts of new and upsetting events at home and abroad. Everything seems to be in a state of flux. Great empires are facing as sudden dissolution or transformation as the neighborhood grocery or the community's time-honored institutions. Enterprises, large or small, local or national, seem destined to feel the impact of forces that cannot be ignored.

We all know, in a general way, what is back of this development. The story has been told so often that its restatement is almost superfluous. The World War, so we are informed, merely accelerated the approach of an inevitable crisis which sooner or later was bound to overtake the modern world. Our machine civilization, with its ever-increasing use of high-powered and automatic production devices, is in violent contrast with the era of handwork. Technological unemployment appears to be with us permanently. Agriculture, continually made more efficient, is likewise tending in the direction of controlled mass production. And so the "technocratic" age is upon us, bringing with it an almost unending series of economic and social problems. The great industrial nations, all striving to be self-contained instead of co-operative, seem to be bent on a life-and-death struggle for raw materials, markets, and colonial outlets for their excess population. In our big cities thousands of unemployed workers are anxiously awaiting the return of prosperity. Governments everywhere are trying to feed their hungry millions and at the same time vainly attempt to balance budgets. Retrenchment at some points and unlimited expenditures at other points have thrown civic and national households into a state of chronic uncertainty. In the meantime, the popular demand for stable conditions and for economic security is approaching the status of a religious dogma. As a corollary, political reorientation is going on in an unprecedented way. While the democratic nations of the western world are experimenting with governmental modifications within the framework of their constitutions, the fascist and collectivistic states are enforcing the dictates of a minority on behalf of an extreme nationalism or a militant proletariat. In this battle of gigantic dimensions the fate of the individual is becoming increasingly insignificant and precarious. Is not the conclusion justified that education, forever concerned with the dignity of individual development and leadership, will be driven either to impotent surrender or to a passionate defense of its most precious convictions and principles?

Internal causes of unrest. Hardly less important than the economic and political factors just mentioned, as causes of our disturbed outlook, are certain subsurface influences. These, too, are well known to any student of modern education. The old landmarks of former days have been rudely shaken. First and foremost, a deified materialistic science must be held largely responsible for an all-too-prevalent mechanistic interpretation of the universe. It is the tragedy of this age that precisely at the time when science is losing its mechanistic bias, the spiritual forces of mankind are held in bondage by a false reverence for alleged scientific finalities that are being discredited by honest scientific thinkers. As yet, psychology, religion, philosophy, and education are suffering from their fatal contact with a machine conception of life and human destiny. Hence the futile attempts to derive valid guiding principles from an experimental philosophy of education which rests on a mechanistic foundation. In proportion as the individual is viewed merely as an inconsequential atom in a relentless evolutionary struggle based on the survival of the fittest, and in proportion as life's ideals are given the status of purely pragmatic and provisional generalizations devoid of intrinsic worth and authority, education is finding it increasingly difficult to rescue its faith in the unique value of the educational process for the individual child. Hence we have a type of mass education which is forced down to the level of the "statistical average." In a world thus robbed of its anchorage and its dependence on clearly charted routes, education is groping about blindly for pilots that appear to know how to steer a course at a safe distance from the rocks of destruction.

How a philosopher views the situation. Under such conditions, how may a teacher preserve sanity of judgment and calmness of mind? More especially, how may we rise above the babel of contemporary voices and rescue for ourselves that unified thinking which is our only escape from hopeless cynicism or personal disintegration? For an answer, let us turn to a philosopher who has been sufficiently close to the popular mood to have developed a deep understanding of our spiritual distemper and its possible correction. Will Durant, in his *Mansions of Philosophy*, has given us one of the most penetrating criticisms of the modern era, together with a revealing diagnosis that might represent a first step in the direction of a cure. Quotations from that work are presented on the following pages.

Human conduct and belief are now undergoing transformations profounder and more disturbing than any since the appearance of wealth and philosophy put an end to the traditional religion of the Greeks. It is the age of Socrates again: our moral life is threatened, and our intellectual life is quickened and enlarged, by the disintegration of ancient customs and beliefs. Everything is new and experimental in our ideas and our actions; nothing is established or certain any more. The rate, complexity, and variety of change in our time are without precedent, even in Periclean days; all forms about us are altered, from the tools that complicate our toil, and the wheels that whirl us restlessly about the earth, to the innovations in our sexual relationships, and the hard disillusionment of our souls. The passage from agriculture to industry, from the village to the town, and from the town to the city, has elevated science, debased art, liberated thought, ended monarchy and aristocracy, generated democracy and socialism, emancipated woman, disrupted marriage, broken down the old moral code, destroyed asceticism with luxuries, replaced Puritanism with Epicureanism, exalted excitement above content, made war less frequent and more terrible, taken from us many of our most cherished religious beliefs, and given us in exchange a mechanical and fatalistic philosophy of life. All things flow, and we are at a loss to find some mooring and stability in the flux.

In every developing civilization a period comes when old instincts and habits prove inadequate to altered stimuli, and ancient institutions and moralities crack like hampering shells under the obstinate growth of life. In one sphere after another, now that we have left the farm and the home for the factory, the office and the world, spontaneous and "natural" modes of order and response break down, and intellect chaotically experiments to replace with conscious guidance the ancestral readiness and simplicity of impulse and wonted ways. Everything must be thought out, from the artificial "formula" with which we feed our children, and the "calories" and "vitamins" of our muddled dietitians, to the bewildered efforts of a revolutionary government to direct and co-ordinate all the haphazard processes of trade. We are like a man who cannot walk without thinking of his legs, or like a player who must analyze every move and stroke as he plays. The happy unity of instinct is gone from us, and we flounder in a sea of reasoning and doubt; in the midst of unprecedented knowledge and power we are uncertain of our purposes, our values, and our goals.

From this confusion the one escape worthy of a mature mind is to rise out of the moment and the part, and contemplate the whole. What we have lost above all is total perspective. Life seems too intricate and mobile for us to grasp its unity and significance; we cease to be citizens and become only individuals; we have no purposes that look beyond our death; we are fragments of men, and nothing more. No one (except Spengler) dares today to survey life in its entirety; analysis leaps and synthesis lags; we fear the experts in every field, and keep ourselves, for safety's sake, lashed to our narrow specialties. Every one knows his part, but is ignorant of its meaning in the play. Life itself grows meaningless, and becomes empty just when it seemed most full.

Let us put aside our fear of inevitable error, and survey all the problems of our state, trying to see each part and puzzle in the light of the whole. We shall define philosophy as total perspective, as mind overspreading life and forging chaos into unity. . . .

For what if we should fatten our purses, or rise to high office, and yet all the while remain ignorantly naïve, coarsely unfurnished in the mind, brutal in behavior, unstable in character, chaotic in desire, and blindly miserable?

Ripeness is all. Perhaps philosophy will give us, if we are faithful to it, a healing unity of soul. We are so slovenly and self-contradictory in our thinking; it may be that we shall clarify ourselves, and pull ourselves together into consistency, and be ashamed to harbor contradictory desires or beliefs. And through this unity of mind may come that unity of purpose and character which makes a personality, and lends some order and dignity to our existence. Philosophy is harmonized knowledge making a harmonious life; it is the self-discipline which lifts us to serenity and freedom. Knowledge is power, but only wisdom is liberty.

Our culture is superficial today, and our knowledge dangerous, because we are rich in mechanisms and poor in purposes. The balance of mind which once came of a warm religious faith is gone; science has taken from us the supernatural bases of our morality, and *all the world seems consumed in a disorderly individualism that reflects the chaotic fragmentation of our character.* We face again the problem that harassed Socrates: how shall we find a natural ethic to replace the supernatural sanctions that have ceased to influence the behavior of men? Without philosophy, without that total vision which unifies purposes and establishes the hierarchy of desires, we fritter away our social heritage in cynical corruption on the one hand, and in revolutionary madness on the other; we abandon in a moment our pacific idealism and plunge into the co-operative suicide of war; we have a hundred thousand politicians, and but a single statesman. *We move about the earth with unprecedented speed, but we do not know, and have not thought, where we are going, or whether we shall find any happiness there for our harassed souls.* We are being destroyed by our knowledge, which has made us drunk with our power. And we shall not be saved without wisdom.[1]

The Crisis in Education

A psychologist looks at education. Having noted the message of the philosopher to our disorganized society, let us turn to a trenchant

[1] Durant, Will, *The Mansions of Philosophy*, pp. vii-ix, x-xi. Simon and Schuster, New York, 1929. The italics in this quotation do not appear in the original text. Italics are introduced here and in all subsequent quotations for the sake of emphasis.

analysis of the educational situation which confronts us at this time. It comes from the pen of a distinguished psychologist who has also given evidence of his keen interest in the deeper meanings of life. Professor Raymond H. Wheeler of the University of Kansas is a leading exponent of configurational psychology. The following abstract from one of his recent monographs is significant here:

Why is our educational system under fire all the way from the kindergarten to our graduate schools? Nearly everybody senses a bad situation; it is doubtful if anybody has completely analyzed what it is. Perhaps the reasons can more adequately be seen if education is deliberately examined as an institution created by the people for the purpose of facilitating social progress. Then the problem requires a close inspection of cultural history, and a clear-cut recognition of a cultural change that is taking place on a world-wide scale, today. This change penetrates deeply into science and is altering the fundamental pattern of our thinking. Indeed, the change is so revolutionary that the conservatively minded person can see nothing in it. No one sees into it whose mind is burdened with inertia, or whose attitude is shaped by nineteenth century points of view.

Essentially, the crisis in education is synchronous with this crisis in the fabric of human culture. Its immediate occasions are numerous. Among them is the discovery that nature has been misconceived, more especially human nature. Science, too, has been misdefined. When man rediscovered science during the Renaissance he overestimated the part played by environment—the stimulus, the objective, the inductive process—as compared with the part played by man himself, in relation to environment. He underestimated the part he played—the dependence of environment on him—in the observation process. He neglected the deductive aspect of experience, itself. Consequently he did not discover until the twentieth century that nature, after all, has a deductive as well as an inductive way about her, implicit in the character of her objective methods. And what is quite as important, it was not until the twentieth century that scientists discovered how beautifully they were letting the deductive aspect of the observation process impose itself upon nature. He [the scientist] had left fact-getting unsafeguarded against errors of logic. He had taken for granted a certain conception of reality and was leaving it uninspected. It was a mechanistic conception.

Meanwhile, it came to be assumed, for purposes of education, that knowledge is more inductive than deductive, almost exclusively inductive. This fatal error has cost us much. The inductive bias assumes knowledge to be built up piecemeal from so much experience here and so much there. The bits are put together. The school curriculum is based upon this thesis from the kindergarten to the graduate school. Separate courses; separate skills. Now we know that knowledge accumulates in no such fashion and that, for the most part, individuals become educated, in spite of, rather than because of, formal efforts to integrate these isolated scraps of knowledge into a single whole, the individual mind. Education has been successful, in a measure, because there are good teachers who have always intuitively sensed the fact that knowledge does not grow in piecemeal fashion. They have been successful teachers because they did not practice the psychology that they had learned, and because they overcame the handicaps under which a divided curriculum placed them. . . .

Culture is unitary, transposing from one end of knowledge to the other, across science, literature, art, philosophy, social codes, theory and practice. It is a single pattern. There is only one mind by which to solve all human problems whether they be in physics, psychology, or politics. And this one mind will solve its problems—if at all—only by beginning, in each case, with identically the same kind of assumptions and by advancing with the same set of methods. Fundamental mistakes made in one field of endeavor will also be made in all the others, at the same time, and for identically the same reasons. The change in human thought, today, is precisely the discovery of such a fundamental error which has transposed across all thought and distorted the nature of all observation. So it is that, today, there is transpiring a universal revolution, encompassing the totality of human thought and attitude. We see it in numerous so-called "movements," the relativistic movement in physical science, the organismic movement in biological science, the configurational movement in psychological science, and the socialistic movement in practical economics and politics. It is no accident that all these events are occurring simultaneously. The revolution is beginning to penetrate into the minds of educators who are beginning to inspect the educational system with the conviction that there is something fundamentally wrong with it, but just what, they do not know. Just what is wrong can best be seen in the light of changes occurring in science itself.

This new conception of nature demands a new conception of education and an altogether different type of teacher training. It demands a widely different type of curriculum than we now have, all the way along the educative scale. It points the way out of the present crisis. It repudiates, unqualifiedly, a mechanistic conception of nature, of evolution, growth and learning. It repudiates the theory that complex things are built up from simple things, that wholes are derived from their parts; it repudiates a mechanistic physics, biology, psychology, a mechanistic conception of the economic and social order, a mechanistic conception of the educative process.

It demands the opposite conception of the learning process than has always been entertained, except by those rare thinkers and teachers of the past whose faithful, intuitive understanding of human nature compelled them to follow their common sense rather than the scientific theories of their day.

This, in general, is the reason why nineteenth century psychology and education, still prevailing, do not yield methods that provide the growing mind with that unity of knowledge which it craves, and makes it useful, nor does it furnish the growing mind with that social insight upon which the integrity of the social order rests. The secret of democracy is co-operation, yet childhood and youth are taught under a scientific philosophy of competition, the philosophy of a mechanistic order. Youth is taught this falsehood in biology, in psychology and in business. . . .

The resistance against new discoveries . . . comes from the man who was once too sure that he was right and now finds it very difficult to change. It comes largely from that leadership in education where one ought to expect the quickest and surest grasp of what is going on in the world today. . . . As it is, *educational leadership is far behind scientific leadership, even far behind the leadership of educated laymen. This leadership in education is so far behind, on the whole, that unless it comes to life, reforms will be accomplished without it. . . .*

Education has spent its substance upon the mechanics of a machine, worrying about technicalities of method and efficiency, without knowledge of what it is that makes the mind efficient. It lacks a philosophy, a goal, an understanding of culture. It is ignorant of history, science, and logic. As a consequence it has set up a shallow educational program, dry and devitalized, that works against the true forces of democracy by stimulating superficiality, conformity and the homogeneity of the mob. Democracy hinges upon unique contributions of the individual, upon individual differences, not upon their elimination. A common ground of knowledge and sympathy can not be obtained or preserved on such a basis. Special skills or special subjects, taught out of relation to culture as a whole, do not integrate; there is no transfer. . . . Methods have the cart before the horse. The true procedure by which we learn is not that of beginning with the "elements" but that of beginning with limited, but unified, general grasps of total situations. Learning is orientation, a single process; it is not several processes put together.

Again, education is on the firing line because it is saturated with a hypocritical and false utilitarianism. Pragmatism has been misdefined. A fact is not true because it works; it works because it is true. Pragmatism overemphasizes means at the expense of ends, method at the expense of purpose, the practical at the expense of the ideal. The pragmatic is that which functions as means toward the ideal, which is the end. There can be no practical out of relation to the ideal, no process out of relation to an end, no progress without the anticipation of a goal which is its own reward. Knowledge should not be taught as subordinate to the practical, but the practical as subordinate to knowledge.

Learning is discovery, always discovery. It is not achieved by mechanical methods of exercise and drill, nor, in the conventional sense of the term, is it subordinate to pain and pleasure. These are symptoms, not causes.

Education has much of its work to do over. That is as it should be. The sciences have had to do it; they are doing it now. When will educators discover what their true task is and build a new temple?[2]

PART TWO

THE STRUGGLE FOR AN ADEQUATE PHILOSOPHY OF EDUCATION[3]

Philosophy and Science as Essential Enterprises

Our constant dependence on philosophy. It has been said that philosophy is an unusually persistent and honest attempt at clear thinking about life's great problems. If such thinking is pursued with sufficient intensity and thoroughness, it will lead inevitably to a consideration of the beginnings and the end of all things. In other words, all true thinking seeks an adequate anchorage or starting point from which it may proceed, and with equal necessity is directed toward a goal. And so, throughout the ages there may be observed the never-ending struggle to find convincing answers concerning such fundamental questions as the following: "Where do we come from?" "Where do we go?" "What is the purpose of all things that exist?" "What is the goal of the universe?" "Why should there be a universe rather than nothing?" "Is anything truly real?" "How can we know?" "What is truth?"

As soon as man emerges from a state of primeval savagery, he begins to ask questions of this sort. To this day, they cannot possibly be evaded. Like a never-ending melody, these questions accompany us from the cradle to the grave. They transcend all of life's activities.

Science and philosophy as complements. It needs to be repeated again and again that both science and philosophy are permanent and essential human enterprises. There should be no quarrel between them. Each has its characteristic domain. If their respective functions were clearly differentiated, much of the confusion in present-day thinking would be eliminated.

[2] Wheeler, Raymond Holder, "The Crisis in Education," *School and Society*, December 9, 1933, pp. 756-759.

[3] In the preparation of this section, the writer has been greatly assisted by Professor M. Demiashkevich's *Introduction to the Philosophy of Education*, American Book Company, New York, 1935. He acknowledges his indebtedness to this remarkably clear and helpful treatise and to an earlier volume by Professor A. N. Whitehead, entitled *The Aims of Education and Other Essays*, The Macmillan Company, New York, 1929.

Science is interested in the discovery of *facts* and in their systematic study. The scientific method has enabled man to enlarge his dominion over this material world to an extent previously considered impossible. The achievements of modern science read like a fairy tale. By the expert use of observation, of laboratory experimentation, and of mathematical analysis, the scientist has fashioned the tools which have enabled him to transform industry, to relieve human suffering, to place heretofore unknown material comforts within the reach of everybody, to extend the confines of knowledge beyond anything held possible in previous centuries.

But this spectacular success of the scientific method has led certain men of science, though usually not the greatest scientists, to look with contempt at "mere speculation." Such an attitude ignores a very simple but all-important distinction between the relative functions of science and philosophy. It is obvious that the scientist can never subject to observation or experimentation either the beginnings of things or their final fate. That is, the question of "ultimates" is beyond his province. *Philosophy begins where science ends.* A scientist may also be a philosopher. But when he speaks as a philosopher, he moves in a different realm.

Science—in the sense of experimental science or "science proper"—*is interested in knowing facts, their immediate origin, their fruits and consequences.* It controls and verifies its own knowledge only within the realm of those facts. *Experimental science does not study—nor can it study—the first causes and last ends of all existence.* Those things cannot, on account of their very nature, be demonstrated in a laboratory. Similarly, experimental science is not concerned with the problem of the last foundations of all knowledge, of the validity of knowledge in general. . . .

All told, experimental science on account of its very nature, resulting from its method of investigation and verification, cannot penetrate below the surface of phenomena or the reality as it appears to be. But humanity clamors for an answer to the question as to what is the ultimate sense of all existence. The history of philosophy, in its metaphysical or ontological doctrines, presents a series of answers for the benefit of all who wish to be assisted in their quest for a rational ultimate certainty by the proved findings and suggestions of great philosophers. Such is the first fundamental fact about the history of philosophy.[4]

It is clear that neither science nor philosophy can furnish *compelling* answers to ultimate questions. Each is moving on a curve

[4] Demiashkevich, *op. cit.,* [3], pp. 43-44, by permission of The Macmillan Company, publishers.

that approaches, but never reaches, its asymptotes. This gap between reality and its human interpretation can be bridged only by enlightened religious faith. It remains true that "where there is no vision, the people perish." *Vision* is the fundamental concern of philosophy and religion.

Education and philosophy. Obviously, education rests on the triple foundation of science, philosophy, and religion. Some educational leaders seem to have forgotten this fact. In their desire to make education purely "objective" and "scientific," they have been misled into an abandonment of philosophic orientation. Here is one of the principal sources of the present confusion in our educational aims, and in our educational theories and programs.

In his Inglis Lecture on *The Way Out of Educational Confusion*, Professor Dewey stated, with perfect truth, that this confusion, which is only too apparent, "is due ultimately to *aimlessness*."[5] Precisely, but what caused this aimlessness? Is it not due in large measure to the very men who are now deploring its arrival? Was it not the same Professor Dewey who urged us *"to abandon the search for absolute and immutable reality and value,"*[6] and who identified religion with superstition?[7] We are certainly facing a peculiar situation when another well-known educator assures us that until "the people" have decided what they want in education, there is no hope for improvement. Is not this a declaration of bankruptcy on the part of these same educational leaders to whom "the people" have looked for guidance? Is not our educational crisis due largely to our uncritical espousal of pragmatic thinking which William James defined as *"the attitude of looking away from first things, principles, categories, supposed necessities,* and of looking toward last things, consequences, *facts*"?[8] But how can "last things" be dissociated from "first things"? Does not a goal presuppose a *starting point* and a *direction*? How can one discuss "consequences" without "antecedents"? At best, a pragmatic attitude leads one to "muddle through somehow," and to adopt a policy of momentary expediency.

[5] Dewey, John, *The Way Out of Educational Confusion,* Inglis Lecture, p. 40, Harvard University Press, Cambridge, 1931.

[6] Dewey, John, *The Quest for Certainty,* p. 311, Minton Balch and Co., New York, 1929.

[7] *Ibid.,* p. 10.

[8] James, William, *Pragmatism,* pp. 54-55, Longmans, Green and Co., New York, 1907.

Instead, it is here contended that the first step in the direction of educational recovery is an honest examination of the "first things and principles" from which we have been turned away to our detriment.

Basic Philosophic Orientation

Importance of basic concepts and assumptions. Every well-informed teacher of mathematics knows the rôle which is occupied in the study of mathematics by its underlying concepts and assumptions or postulates. The first genuine example of a postulational system was Euclid's famous textbook on geometry.[9] It taught the world how to organize knowledge on a basis of definitions and first principles. We have ample reason for saying that the deductive techniques of mathematics have furnished a pattern that must eventually be followed in every other field of research.[10] Only by being equally careful in its fundamental approach can education hope to become "scientific." The confusion and uncertainty characteristic of education today will be overcome only in proportion as its basic concepts and its guiding principles can be stated in less ambiguous terms. This is impossible without a clear understanding of the working vocabulary of the contributing sciences and of the great historic systems of philosophy.

In particular, the educator should be familiar with the meaning and philosophic import of such terms as materialism (naturalism, realism), idealism, skepticism, dualism, pluralism, pragmatism, and the like. He should know that each complete system of philosophy has its own *metaphysics*, or conception of the ultimate reality, its own *epistemology*, or theory of knowledge, and its own *ethics*, or doctrine of conduct and values.[11]

Thus, *the materialistic approach* to the problem of reality is *empiricism*. Materialism asserts that all knowledge is based on the experience of our senses. Ethically, materialism leads to *determinism*, according to which man's conduct is governed by mechanical natural laws, the inherent properties of matter. Hence, a con-

[9] Keyser, C. J., *Thinking about Thinking*, p. 25, E. P. Dutton and Co., New York, 1926.

[10] *Ibid.*, p. 27. See also Bell, E. T., *The Search for Truth*, The Williams & Wilkins Co., Baltimore, 1934.

[11] See Demiashkevich, *op. cit.* [3], Chaps. I-III; see also Patrick, G. T. W., *Introduction to Philosophy*, Chaps. I-IV, Houghton Mifflin Co., Boston, 1924, revised, 1935.

sistent materialist must deny the freedom of the will as well as moral responsibility.

Idealism, on the other hand, asserts that the ultimate reality is *spiritual* or immaterial. Its method of obtaining knowledge of this true reality is the speculation of our *reason.* Ethically, idealism asserts the *freedom of the will.*

Just as we now have a multiplicity of consistent systems of mathematical axioms, so we may approach life's ultimate problems in the spirit of one or the other of the great philosophic systems. And *choose we must.* In the following passage, Professor Ralph Beatley has stated this thought very clearly, though perhaps the negative form of his opening sentences may seem unwarranted:

> We can never fully know what is the true end of man; we can never know the precise purpose of the secondary school; we can never know the part which mathematics ought to play in training the youth of this nation. These matters rest ultimately upon philosophy, and there are many philosophies. *The answers vary according to the philosophy we choose.* If we choose that one which yields the most satisfactory answers, we must realize that this very satisfaction, however obtained —from standardized tests, subjective opinion, the experience of the ages, common sense—derives its ultimate validity for us from some philosophy. Our choice of answer, of what to us is satisfying, reflects indeed our choice of *a* philosophy. Our neighbor may choose differently. As between the respective merits of these philosophies, who can arbitrate?
>
> Even if we can never surely know, a common philosophical notion prompts us to the belief that if we would live, be good citizens and good teachers, we must try *to discover by thinking the most satisfying philosophy concerning the education of boys and girls in secondary schools.*[12]

In the light of these preliminary considerations, we shall now examine, however briefly and imperfectly, the dominant educational philosophies of our day. It will appear that our main task is a synthesis that avoids extreme positions and fallacious inferences.

The Philosophy of Pragmatic Instrumentalism

Preliminary statement. The attempt to summarize in a few pages the educational implications of so complex a philosophic movement as is represented by the various phases of pragmatic thinking which

[12] Beatley, Ralph, "Coherence and Diversity in Secondary Mathematics," *The Eighth Yearbook, National Council of Teachers of Mathematics,* p. 165, Bureau of Publications, Teachers College, Columbia University, New York, 1933.

are now current constitutes an almost insuperable task. But since it is undeniable that the influence of pragmatism, particularly in the form given it by Professor Dewey, has been an all-pervasive educational ferment of our generation, such analysis, however fragmentary, cannot be evaded. The very name of the movement carries us back to ancient Greece. The term "pragmatic" was employed by Kant in one of his most important treatises. But it was William James, through his volume on *Pragmatism*, who gave general currency to this mode of thinking. We are here concerned primarily with that phase of the pragmatic attitude which has become known as *instrumentalism.* Its chief exponent in this country is John Dewey, who has given us his views on the subject in an impressive series of volumes or monographs.[13] Unfortunately, Professor Dewey's style is difficult and his English often seems impenetrable. As a result, it is doubtful if many teachers have ever had the time or the energy to read one of Dewey's major works. Most of them have been affected by Dewey's educational ideas only indirectly, through a large number of slogans. It is these slogans, repeated uncritically until they have been accepted as axioms, which have been responsible, in the main, for the educational influence of Professor Dewey and his numerous followers.

Not long ago, nearly every appraisal of Dewey's educational contributions exhibited a spirit of frenzied enthusiasm bordering on idolatry.[14] A more conservative estimate is now emerging, and the baneful effects of pragmatic thinking are at last beginning to be realized.

Pragmatism as a method. We now have a number of excellent critical discussions of pragmatism, and particularly of Dewey's instrumentalism. In the opinion of the writer, nothing would pay greater educational dividends to the teaching profession just now than a careful study of these candid studies.[15]

[13] See especially *Democracy and Education, Reconstruction in Philosophy, Human Nature and Conduct, Experience and Nature, The Quest for Certainty,* and *Art as Experience.* The series of essays entitled *Creative Intelligence*, by Dewey and his followers, is also of importance.

[14] For illustrations, the reader may be referred to the Preface of Professor H. H. Horne's commentary on Dewey's *Democracy and Education*, entitled *The Democratic Philosophy of Education*, pp. ix ff., The Macmillan Company, New York, 1932.

[15] In addition to Horne's excellent commentary, reference should here be made to the following contributions: Demiashkevich, M., *op. cit.*, pp. 107-117; Horne, H. H., *This New Education,* Abingdon Press, New York, 1931; Kandel, I. L., "The

The manifold-hued nature of pragmatism is characterized by Patrick as follows:

Pragmatism is a tendency and a movement rather than a philosophy. In fact it holds philosophical systems in profound suspicion. It is more like a "corridor" through which one may enter upon philosophical studies. *It is an attitude and a habit of thought—a habit of looking forward to results rather than backward to first principles. Everything is to be judged by its fruits, by its consequences. Thus it follows that any idea, theory, or dispute which does not make a difference in its practical consequences for us ceases at once to have any significance.* All these are simply dropped; they cannot be tested. Hence a great number of ancient philosophical controversies, theories, hypotheses, systems just collapse; they fade away under this rigid pragmatic test. . . .

In the older philosophy there was much talk about certain ideas, such as God, Matter, Reason, the Absolute, the Soul. These ideas were ultimate, and we felt that we could rest in them. But the Pragmatist does not take this attitude toward them. *He does not want to rest. He inquires as to their cash value. He will put them to work and see what consequences they may yield. If they will not work, they are not true.* Pragmatism unstiffens all our theories, limbers them up and sets each one at work. In actual life we have always to deal with definite concrete situations, and these situations are to be met and solved on their own merits—not on abstract traditional principles. *Life is a maze through which we are threading our way as best we can, finding the path as we go along. Answers which solved former situations will not solve this one. Everything changes, grows, develops; nothing is fixed, static, final.*

Even moral laws change; they grow and become perfected. *There are no fixed or final moral laws and no eternal principles either of conduct or knowledge. Reality is in the making; you and I are making it.* The road to the future is an open road, obstructed by *no overruling providence or limiting fates, and determined by no* a priori *principles of thought. Reality is found in the flow of experience. The world is moving toward no predetermined end;* each hill is surmounted as it comes into view. What happens next is not determined, but is contingent upon what has happened. *Life is a series of problems to be solved*—a succession of real struggles with real difficulties. *To think is to deal effectively with these problems—and ideas are tools to help in the solution.*

Reality is fluid, changing, evolving. Pictures of a God-made, perfect world, governed by eternal principles of justice or by eternal math-

---

Philosophy Underlying the System of Education in the United States," *The Educational Yearbook, 1929,* International Institute of Teachers College, Columbia University; Kilpatrick, W. H., "Tendencies in Educational Philosophy," Chap. III in *Twenty-five Years of Education* (I. L. Kandel, Editor), The Macmillan Co., New York, 1924.

ematical laws, are just fancies. Leibniz' theory of the world as the best of all possible worlds is false. *The only real world is the world of real experience. . . .*

The spirit of Pragmatism is the spirit of youth, adventure, and experimentation; it has no patience with idle vaporings about fate and destiny. *No philosophical ideas are true which cannot be put to some practical use. Take such words as God, free-will, or design. Other than practical significance, says James, they have none.*[16]

In other words, pragmatism is only "a manner of *approach* to philosophy, a doctrine of logical *method,* and not a theory of the Universe." According to the pragmatist, "matter and spirit, body and soul, subject and object, *a priori* rules of thought or conduct —all these are too far away, too abstract, too unreal." Hence "philosophy cannot begin with them; it has to do only with experience; *the world is a world of pure experience.*"

Now, *what is experience?* The instrumentalist defines it, incompletely, as "the interaction of the living organism with its physical and social environment." Again, "ideas are not psychical entities or subjective representations of an objective reality; they are *plans of action,* taking into account future consequences with reference to the weal or woe of the organism. *By intelligence is meant just this ability to organize responses with constant reference to future consequences.*"[17]

A Critical Appraisal of Pragmatic Instrumentalism

We have gone far enough in our analysis to make clear some of the cardinal flaws in this mode of approach.

1. To say that "all experiences are real," is one thing. But to affirm also the truth of the *converse* of this statement leads to grotesque absurdities. How can one assert that "all realities are experiences?" Certainly historical events are real, though they can no longer be "experienced" by any living organism. All race experience is real, but no individual could ever "reconstruct" all of it, even though he lived a million years. Such ethical ideals as justice, truth, and honesty are real, whether we appreciate their existence personally or not. "The school is life," we are told. But the school can never reproduce or "reconstruct" *all* of life. We can never eliminate from the scene the equally important *vicarious* experience of the race.

[16] Patrick, G. T. W., *op. cit.* [11], pp. 378 ff. (The bracketed figure here, and in subsequent notes, represents the title cited in a preceding footnote indicated by that number.) [17] *Ibid.*, p. 386.

2. The pragmatist's conception of truth, as ordinarily stated, leads to complete intellectual and ethical nihilism. We are told that "truth is that which works," and that "the truth of an idea is not a stagnant property inherent in it." Instead, "truth *happens* to an idea; it *becomes* true, is *made* true by events!"[18] Now, as Patrick correctly points out, "the *verification* or *test* of truth is one thing, while the *structure* of truth is quite another."[19]

3. While it is often very helpful to look upon concepts, ideas, and thinking as instruments or tools for solving life's problems, *it is not true that all knowledge and all thinking are instrumental.* Meditation, contemplation, reflection, philosophic speculation, and the like, are certainly of vast importance, even though they are at their best when not concerned with utilitarian considerations or practical "consequences."

4. It may stimulate us to think of the world as "constantly in the making," of growth leading to more growth, and still more growth, by experimental living," and the like. But what is the incentive for constant striving, for all this experimentation, without a *purpose*, without a worthy and inspiring *goal*? How can growth lead to *desirable* further growth without some guiding principle? *The absence of such standards, however provisional they may be, must inevitably lead to social and cultural chaos.*

These inescapable implications of the pragmatic attitude have often been overlooked by Professor Dewey's many admirers. Education has paid a heavy price for these weaknesses in his theory. Only a radical revision of our educational creeds can bring order out of the present confusion and aimlessness which Professor Dewey himself deplores, but for which his mode of thinking is so largely responsible.

THE PHILOSOPHY OF CHANGE AND OF SOCIAL RECONSTRUCTION

The concept of "change." "Everything flows," said Heraclitus, some five hundred years before the Christian era. Since then, the concept of change, of flux, has continuously affected philosophic discussions. Today, the apostles of change seem anxious to make it the dominant concept of education. Just at present they represent a powerful pressure group, and on the surface their arguments seem both convincing and impressive.

[18] *Ibid.*, p. 391.
[19] *Ibid.*, p. 393.

In the first place, the doctrine of evolution, so we are told, clearly points out the ceaseless transformation that has been going on in the *physical* world. From the tiniest atom to the island universes floating in outermost space, nothing seems immune from the impact that constantly modifies the cosmic scene.

The same appears to be true of all *human institutions*, as Spengler has shown. Culture epochs come and go, and civilizations rise and fall, in obedience to the same inexorable laws of growth and decay that govern all biological organisms.

In this evolutionary setting, it has become our fate to experience the arrival of the *machine age* with its unavoidable outcome, the *industrial revolution.* As a result the time-honored customs and beliefs of society are being subjected to transforming influences beyond the wildest dreams of even a generation ago. The whole world is being tested in a crucible of fire. Will the result be a new birth or complete chaos?

The doctrine of change in the field of education. The school, so we are told, cannot remain indifferent to the titanic conflict we are witnessing from day to day. In ringing phrases, the militant educational reformer tells us that the school has degenerated into a museum of fossilized procedures, of antiquated curricula and objectives. We need a complete turnover. Above all, the school must assist in the "continuous reconstruction of society" and in bringing about the adjustments demanded by the new economic order. This thesis has been stated as follows by Professor Dewey:

> An identity, an equation, exists between the urgent social need of the present and that of education. Society, in order to solve its own problems and remedy its own ills, needs to employ science and technology for social instead of merely private ends. *This need for a society in which experimental inquiry and planning for social ends are organically contained is also the need for a new education.* In one case as in the other, there is supplied a new dynamic in conduct and there is required *the co-operative use of intelligence* on a social scale in behalf of *social values.*[20]

More recently Professor Kilpatrick has further expressed his views on the same subject:

> Most inclusive of the new developments now demanding attention is the fact of modern rapid change, much discussed but still tragically disregarded in social thinking and educational practise.

[20] Kilpatrick, William H. (Editor), *The Educational Frontier,* p. 64, D. Appleton-Century Co., New York, 1933.

A modern notion of change has emerged. Affairs develop in ever novel fashion. New situations continually confront. New aims arise. Old knowledge and habits are reworked in with the new conditions, and new results appear. Culture thus accumulates: ever new knowledge, distinctions, attitudes, and technics. Efficiency thus increases and social intelligence grows. Individual intelligence sharing the new cultural product should grow correlatively.

Amid ever novel conditions, thinking is stressed, mere habit could not suffice. Each new situation is a problem, demanding its study and thought. We try out our best thought plan; we watch whether it works. Each new program is thus an experiment. Amid changing conditions, we live experimentally, must do so. Education ceases then to be mere acquisition of something handed down. It too becomes experimental. Otherwise, it were no adequate preparation for a changing and experimental life.

In a rapidly changing civilization new social problems thus continually arise, *with ever new solutions proposed.* These new solutions, democracy demands, must be passed upon by the people. Citizens must then be continually studying, criticizing their institutions to improve them. Social education thus must become a lifelong process. This must begin before twenty-one, or the person is sadly handicapped and probably biased against study and intelligent criticism.

The schools must accept the new task. The pupils must learn ever better, with their increasing years, to study and criticize our institutional life, in order, intelligently, to help improve it. *The alternative is unintelligent indoctrination in the status quo.*[21]

Extreme demands. No one should object to a sensible and balanced demand for an "up-to-date" educational program. Our future citizens, taxpayers, and wage earners can hardly learn too much about the conditions which they must face all too soon. It would be highly menacing to our institutions to bring up the rising generation in ignorance of the pressing social, economic, and political problems that surround us on all sides. But does that obvious duty warrant a disregard of all other educational values or needs that we have held sacred thus far? In a frenzy of enthusiasm, the social reformer would start with a clean slate. He has but one objective, to make the school *"society-centered."* In the interest of this new gospel, everything else must be sacrificed. Such hoary subjects as Latin and mathematics must go, foreign languages are taboo, science is tolerated for its "social service values," and literature and art are retained as leisure-time pastimes. Any defense of the "old-line subjects" is denounced as the outcry of "vested interests."

[21] Kilpatrick, William H., "New Developments, New Demands," *The Journal of the National Education Association,* November, 1935, p. 261.

Apparently the pupils will be expected to give nearly all their time to current political and *economic* problems. They are to learn about unemployment and its causes, about technocracy and the "economy of abundance," about taxation, production costs, wages, and profits. Are they also to be informed about the real or alleged dangers of capitalism, about the merits and possible flaws of socialism, communism, and fascism?

One gets the impression that under this new dispensation of social reconstruction, each school is to be transformed into a continuous debating society. The schoolroom is to be kept in constant touch with the march of events by every available means, such as excursions, visits to factories, bulletins, news letters, movies, and radio talks. Newspapers and periodicals are to replace the textbooks. Intricate and highly "explosive" problems that have baffled the ingenuity of an army of experts are presently to be attacked by immature boys and girls, many of whom cannot read seventh grade books, cannot write an English sentence, and cannot perform the simplest computations.

A college professor has remarked that if all the economists and statisticians of the world were placed in a line, end to end, they would not reach a conclusion. Perhaps that is the reason why the reformer expects such great results from pupils who are "unburdened by the useless knowledges and skills of the traditional curricula." That many of them are thus unburdened, every teacher knows only too well.[22]

A saner attitude. Significantly enough, the teachers who are to be the principal agents in creating the new, society-centered school, are beginning to realize some of the difficulties that lie ahead of them. They know only too well that whenever the social studies still appear on the educational menu under such specialized designations as history, civics, and economics, the "average pupil" extends to them the same impartial sales resistance that all the other "vested interests" of the school have so long been aware of. These teachers are beginning to suspect that the trick of using a new name for these "subjects" is not going to make a new being out of a poorly

[22] For a survey of the extreme views suggested above the reader may be referred to Everett, S. (Editor), *A Challenge to Secondary Education,* D. Appleton-Century Co., New York, 1935. For a contrasting point of view, see Demiashkevich, M., "The Algebra of Civics," *The Harvard Teachers Record,* February, 1935; and Demiashkevich, M., *Introduction to the Philosophy of Education,* pp. 197 ff., American Book Company, 1935.

prepared, aimless youngster. In their group discussions, they point out the hopelessness of assigning to the teachers of the social studies the rôle of super engineers of the new educational age.

It is encouraging to observe this sane orientation in the authoritative pronouncements that have recently been issued by the Commission on the Social Studies. In an impressive series of reports and studies, some of which are still in preparation, the Commission is submitting its findings and suggestions. Of particular interest is the volume entitled *Conclusions and Recommendations.*[23] Chapter Two of this volume presents a "Frame of Reference"; Chapter Three offers a discussion of "Philosophy and Purpose in Education"; Chapter Four is concerned with the "Selection and Organization of Materials of Instruction." Much of this report should be brought to the attention of every teacher, and particularly to the attention of every iconoclast who, in the name of the "social studies," would demolish the entire educational structure.

We submit merely the following statements as samples of the commendable sanity and thoroughness which, on the whole, characterize the work of the Commission on the Social Studies:

The main function of the social sciences is the acquisition of accurate knowledge of, and informed insight into, man and society; that of social science instruction is the transmission of such knowledge and insight, with attendant skills and loyalties, to the individuals composing society. Regardless of the special circumstances of a given time, these functions are vitally important and likely to be effective in the measure of the breadth and depth of their conception, involving a real knowledge of man and society under most diverse conditions and circumstances.

. . . Scholarship has its own imperatives, and to say that science exists merely to serve the instant need of things, causes, or parties is to betray a fatal ignorance of its nature and of inexorable movements in thought.

The Commission believes that fundamentally the disinterested pursuit of truth and the permanent interests of society as a whole are not, and cannot be, incompatible, and that both the social scientist in his study and the teacher of any social science in his classroom are committed to scholarly, scientific ideals inherent in their profession and occupation.[24]

[23] *Report of the Commission on the Social Studies,* Charles Scribner's Sons, New York, 1934. For a review of the final volume, *Conclusions and Recommendations,* see *The Social Studies,* Vol. XXV, October, 1934, especially the paper by Mr. Kenneth E. Gell, pp. 288-293.

[24] *Ibid.,* pp. 7-8.

The Philosophy of Permanent Values

The concept of "invariance." Every student of mathematics has occasion to learn that this great science is not exclusively concerned with "variables" and with "transformations." It is quite as much interested in "constants," so much so that the concept of *invariance* has become one of the key ideas of mathematics. Thus, the elementary number facts will always remain the same. Two and two will always be four. The sum of the angles of a triangle, in Euclidean geometry, will always be 180°. The quadratic formula has an unchanging relation to the standard quadratic equation. This idea of invariance, as the mathematician sees it, was given currency especially by Professor Felix Klein, and it has been discussed admirably by many other writers.[25]

Now, the pragmatist seems constitutionally unable to tolerate this idea of permanence, of enduring backgrounds, of controlling "frames of reference." He feels that he is being put into a strait-jacket by any attempt at "indoctrination," at binding agreements, at fixed principles of action or of conduct. His style is cramped, so to speak, by anything but a provisional and highly "experimental" attitude toward reality. He desires to be absolutely free, "reconstructing" the universe as he sees fit. He is forever on the road toward ever-shifting horizons. To a certain extent, this has come to be a national characteristic. Like many summer tourists, we are "always going full blast—nowhere."

But this old world of ours has been a long time in the making. Somehow, after billions of years of cosmic adventure, it seems to have achieved a few settled adjustments. If this had not been the case, there would be no super galaxies, and certainly no solar *system.* There would be no biological organisms, for the very essence of life is *organization*, and that implies a *plan*. When man arrived on the scene, he found no alternative but that of "accepting the universe," its law of gravitation, its laws of health, and—if he was wise—its laws or "commandments" of co-operative effort.

Education and its frames of reference. All the forces of nature seem to "work together for good," and man, unless he courted destruction, found it to his advantage to conform to this plan. Ob-

[25] See Keyser, C. J., "Mathematics and the Question of Cosmic Mind," *Scripta Mathematica,* New York, 1935; Smith, D. E., "The Poetry of Mathematics and Other Essays," *Scripta Mathematica,* New York, 1934; Bell, E. T., *The Queen of the Sciences,* The Williams and Wilkins Co., Baltimore, 1931.

viously, then, there could be no "experience," and hence no education, without the interaction of *planful* organisms and a *planned* physical world. An aimless chaos is hardly a fit laboratory for "experience," for it lacks the very first presuppositions of experience, which are dependability, order, and duration. The universe, we are reminded by Sir James Jeans, appears to be a Great Thought, the work of a Supreme Architect, a Master Mathematician.[26]

Have not many of our educators forgotten the facts of invariance in their personal and social philosophy? Fortunately, the revolt against complacent agnosticism or skepticism is steadily growing.[27] Thus, Professor T. V. Smith of the University of Chicago has rendered a real service by analyzing the relation between skepticism and enduring values.[28] In a chapter devoted to an appreciative study of the late Justice Holmes, there occurs this passage:

> All men accept the universe. "Egad, they'd better!" True, not all accept it, as did Holmes, like a gentleman. Some take it lying down, resigned to whatever comes. Some take it with raised voice, defying gods and devils. Some take it dumbly, querying neither whence nor whither. Some take it wide-eyed, a-wonder at all that is, a-twitter at what may be. Some take it equivocally, today aflame, tomorrow hardly a flicker, day after tomorrow dumb despair.[29]

So it comes down to this: Shall we assist our boys and girls in accepting the universe *intelligently* and "like a gentleman," or in the spirit of "dumb despair"? For there is no doubt that radical empiricism, as Bertrand Russell frankly admits in his own case,

[26] Jeans, Sir James, "The Universe A Great Thought," Chap. XIV of *Has Science Discovered God?* (Edward H. Cotton, Editor), Thomas Y. Crowell Co., New York, 1931. See also Jeans, *The Universe Around Us* (1929, revised 1934), and *The Mysterious Universe* (1930, revised 1932), The Macmillan Co., New York.

[27] Admirable expositions of the attitude of leading scientists toward metaphysical or religious problems are now available. See, for example, "A Symposium of Modern Scientific Opinion," gathered and edited by Edward H. Cotton, and published under the title *Has Science Discovered God?* cited above [26]. Among the scientists represented are Millikan, Einstein, Julian S. Huxley, Sir J. Arthur Thomson, Michael Pupin, John Langdon-Davies, Sir James Jeans, Sir Oliver Lodge. See also Carrel, Alexis, *Man, the Unknown,* Harper and Brothers, New York, 1935; Eddington, Arthur S., *The Nature of the Physical World,* The Macmillan Co., New York, 1928; Eddington, Arthur S., *Science and the Unseen World,* The Macmillan Co., New York, 1929; Eddington, Arthur S., *New Pathways in Science,* The Macmillan Co., New York, 1934; and, especially Mason, Frances (Editor), *The Great Design,* The Macmillan Company, New York, 1935.

[28] Smith, T. V., *Creative Sceptics,* Willett, Clark & Company, Chicago, 1934.

[29] *Ibid.,* p. 189.

can only lead to "the debris of a universe in ruins" and hence to the dubious "foundation of unyielding despair."[30]

"Everything in nature is engaged in writing its own history," said Emerson. But every real story is based on a plan, a plot. *Shall we not give to our young people enough orientation and perspective to enable them to select wisely their own life plan, within the framework of permanent backgrounds?*

Wanted—A Comprehensive and Consistent Educational Philosophy

Three permanent centers of interest. It is the function of education to assist growing human beings in developing satisfactory relationships between themselves and the world into which they are born. This process is obviously an endless one, and it has many aspects. When it is directed too exclusively to only one or two of its legitimate objectives, it becomes one-sided and, to that extent, a caricature of what it should be. This is precisely what has happened again and again.

1. Educational programs which are concerned primarily with the real or alleged needs of the "educand," the individual to be educated, lead to *"child-centered"* schools. This type of orientation is particularly marked in the modern school.

2. When the educational process is built principally around the real or supposed demands of the educand's *social group*, or the *community*, we have a *"society-centered"* program of education. We know that this type of emphasis has been a concern of education since the days of primitive man.

3. In so far as both the individual and society are ultimately governed by *cosmic* processes and by such forces as account for the *historic* evolution of human civilization and of human institutions, education cannot ignore the objective consideration of these *permanent* causal factors and their underlying relationships. In its extreme form, such an emphasis leads to a *"curriculum-centered"* school.

Importance of a balanced educational orientation. It is essential that the educator should never lose sight of the *three* poles around which his work will always have to rotate. If he forgets or ignores any one of them, he will eventually wreck the educational

[30] Russell, Bertrand, *Mysticism and Life and Other Essays*, pp. 47 ff., Longmans, Green and Co., New York, 1921.

machine. Our "child-centered" schools and the "activity program" only too clearly show the consequences of a narrowly *subjective* attitude. Again, a purely *social* emphasis leads to a type of regimentation and autocratic indoctrination that we now witness in certain European and Asiatic countries. Finally, the mechanized transmission of the accumulated wisdom of the ages invites the danger of excessive formalism and a resulting decay of mental and spiritual creativeness.

Is it not possible to give due attention to all three of these essential aspects of the educational process? Can we not avoid a constant swerving from one extreme to the other? By so doing, we shall at last eliminate the glaring inconsistencies in our current educational philosophies. Some of these inconsistencies will receive further attention in later sections of this discussion.

The present struggle for a dynamic philosophy of education in a changing world needs to be oriented, above all, by a consideration of *ultimate objectives*, and not merely of immediate interests. In the language of Professor G. T. W. Patrick:

> The great things of the world have been done by men who were inspired by *great ideals*, ideals of justice, righteousness, beauty, and truth. *These lofty ideals are not something to be made and then tested by their satisfactoriness; they are something to be attained.* Beauty which exists just to be appreciated, truth which exists just to be contemplated, laws of nature which just have to be discovered and wondered at, ideals which just have to be aspired to—all these great things would seem to have no place in pragmatic philosophy, which is too subjective. *Something eternal must draw us on.*[31]

PART THREE

THE PRESENT STATUS OF "PROGRESSIVE" EDUCATION

Introductory statement. What is now known as "progressive" education, or as the "new" education, is the culmination of a long development. At first very sporadic and of uncertain character, the movement has now acquired the zest of a cult. In its extreme form, it has assumed an uncompromising intolerance that aims to destroy the traditional educational program. Its impact on school procedures, on curricula and standards, may be observed at every turn. To teachers of mathematics this movement should be of considerable interest, since it has given approval to the curious idea that

[31] Patrick, G. T. W., *op. cit.*, [11], p. 397.

"mathematics is useless to the average child and therefore must either be eliminated or restricted to its few practical and incidental functions." The threatened removal of mathematics from our high school curricula, for the majority of secondary pupils, is due in large measure to propaganda circulated by the uninformed.

It will be shown in the pages that follow that "progressive" education is based on a rather involved philosophy of education. Many of its aspects are contradictory. Like every other pronounced departure from the beaten path, it has produced a conservative and a radical group of devotees. It is the extremism of the latter group that is forcing into the open a critical examination of the fundamental assumptions which govern the "activity movement" and the programs of the "progressive" schools.

The Activity Movement

A brief retrospect. The *Thirty-third Yearbook of the National Society for the Study of Education* may be regarded as the best single source of information now available on the activity movement.[32] Its second chapter (pp. 19-43) offers a "historical sketch of activism," contributed by Professor Thomas Woody, which shows how by gradual steps man made his way from the authoritarian, passive type of education imposed by the clan, or the community, or the printed page, "to the authority of nature." *"Recognizing law in man's nature, educators came to dream of harmonizing his education with that law. Herein is to be found the source of modern activism."* Professor Woody presents an informing summary of the evolution of this idea through the centuries, especially since the days of the Renaissance and the Reformation. Outstanding supporters of the naturalistic trend in education, thus interpreted, were Comenius, Rousseau, Pestalozzi, Froebel, and Herbart. The American phase of the movement dates back more than a century. We are reminded that the designation, "the new education," has been "part of the American pedagogical vocabulary for a hundred years; the ideas, if not the words, have been current in Western European writing to some extent for five hundred years, and to a pronounced extent for three hundred." But it was chiefly the philosophy and the psychology of Professor Dewey that launched activism in a new highly dynamic form and with a revolutionary

[32] "The Activity Movement," *The Thirty-third Yearbook of the National Society for the Study of Education,* Public School Publishing Company, Bloomington, Ill., 1934.

social implication and momentum. Dewey not merely epitomized the contributions of his many predecessors. He and his followers transformed and greatly intensified the central concepts of activism with the aid of more recent psychological and sociological findings or assumptions.

Characteristics of the activity movement. In the third chapter of the yearbook mentioned above, Professor Kilpatrick attempts the considerable task of formulating a coherent "definition of the activity movement today." He states that in their effort "to find an authentic picture," the Yearbook Committee "collected and studied (1) forty-two expert-made definitions; (2) twenty-five carefully selected published curricula, illustrating the activity program; and (3) fifteen books giving authoritative treatment of the subject." These forty-two definitions are given in Appendix 1, while the courses of study and the fifteen books are listed in later appendices. Professor Kilpatrick's summary occupies nearly twenty pages. It records a large variety of opinions and trends with regard to thirteen major aspects of the educational process. As to ways of interpreting the principle of activity in actual school work, Chapter IV of the same yearbook, prepared by four other reviewers, lists "six groups of practices" arranged in ascending order, from a moderate and merely incidental use of activities, experiments, demonstrations, pupil participation, excursions, and the like, to "so profound an acceptance of the faith that the learner develops through his own initiated activity that there are distrust of guidance, lest it transgress individual possibilities, and great emphasis upon study of the individual and upon helping him to further his own efforts." As to the concept of "activity," Chapter IV offers the following statements in final summary:

> Central among the many meanings that the proponents of the activity principle have in mind are kinds of work (1) that enlist the personal concern of the learner in what he is doing, (2) that involve participation of the learner in the life about him, (3) that encourage the learner to initiate action that will further the things in which he engages, (4) that assume and teach personal responsibility for the consequences of one's own doing, (5) that foster creative self-expression as a means and a manifestation of the developing self, (6) that deal with the learner's reality and endeavor to teach the learner to face his own reality, and (7) that assume the necessity of a freedom which makes possible this dynamic living on the part of the learner. Some proponents seem to use the term "activity" as a brief way of postulat-

ing these elements in the learning program. They assume that the term "activity program" implies all these characteristics.[33]

Dean Clyde Hissong, in his study of the activity movement, lists certain basic ideas or principles which the majority of the schools concerned seem to endorse.[34] He gives particular prominence to such items as (1) good health; (2) "learning by doing"; (3) an atmosphere of freedom; (4) the avoidance of standardization and tests in favor of individual development; (5) social adjustment and extensive co-operation; (6) opportunity for creative expression; (7) joyous and creative learning; (8) the attempt to satisfy the pupils' "needs" rather than following a prescribed curriculum.

From these and many similar studies the inference may readily be derived that all good schools could at once subscribe to a large part of the activity program. An inspection of published courses of study and of booklets which are based on this creed usually reveals an unsuspected wealth of excellent materials of instruction, of new means of motivation, and of opportunities for "joyous participation" and creative self-expression, such as any real teacher should welcome.[35]

The "danger zones" of the movement, and they are very real, are caused mainly by (1) an inaccurate conception of the learning process; (2) an unsound doctrine of freedom; (3) a disregard of standards and of critical thinking; (4) an unwarranted neglect of organized knowledge and of the essential tools of learning. Unless and until these defects are corrected, the activity movement must be regarded with caution by all true friends of American education.

Weak spots of activism. The activist asserts that we *"learn by doing."* To what extent is this true?

Now, biologists and psychologists have made us thoroughly familiar with the fact that "organic behavior is determined by the structure of the organism." More than that, it now sounds like a truism that "a child learns to do by doing, as does any other organism."[36] Thus, we learn to walk only by actual walking, to speak and sing

[33] *Ibid.*, pp. 65-66.

[34] Hissong, Clyde, *The Activity Movement*, Warwick & York, Baltimore, 1932.

[35] For illustrations of this type, see, for example, the pamphlet entitled *Cardinal Objectives in Elementary Education*, issued by the University of the State of New York, Albany, 1929.

[36] See Gray, J. Stanley, "A Biological View of Behavior Modification," in the *Journal of Educational Psychology*, November, 1932, for an able discussion of the biological aspects of behavior.

only by using our vocal chords, and to make a multitude of necessary adjustments only by the corresponding bodily responses.

But is it true that all learning implies gross muscular or neurological activities, and in exactly the same sense? Do we learn only by "doing"? To what extent can all the subtle but vastly important mental reactions of the child be covered by this formula? Are thinking, reflection, appreciation, and the like, to be regarded as "activities"? If so, are these on the same level as playing a game, or cutting out paper dolls? Again, race experience certainly cannot all be "reconstructed" in the school by means of activities, and yet this experience cannot possibly be ignored. Hence the slogan that we "learn by doing" needs to be clarified and to be restricted to its proper domain.

Finally, to what extent do activities, unless they are organized with the greatest care, lead to cumulative growth, or to growth in the right direction? Can the unguided activities of groping children ever produce harmonious educational results?

Let us now review a few significant statements on this subject from educators who cannot be accused of lack of interest in the "new education."

Stanwood Cobb, one of the founders of the Progressive Education Association and at present one of its vice presidents, widely known as the director of the Chevy Chase Country Day School, devoted an entire chapter of a recent volume to "the limitations of activity education."[37] From this chapter we quote the following significant passages:

> The tendency today is not so much that the activity method will not be used by teachers, as that it will be abused by them. As in every reform, there is danger of too great a reaction.
>
> *A common defect in the use of activity projects is the neglect to assure definite cultural results.* . . . In other words, *the activity project should be a means toward definite educational goals and not an end in itself.* . . .
>
> *The tools and techniques of learning—such as reading, writing, and arithmetic—cannot be learned by the project method but only motivated by it.* These skills must be made automatic by means of much drill and practice. . . .
>
> *Lazy, unambitious, and slow-temperament children do not respond well to the project method. They learn very little by means of it.*

[37] Cobb, Stanwood, *New Horizons for the Child,* Chap. VIII, The Avalon Press, Washington, 1934.

Such children cannot achieve their best academic results except by means of academic pressure and discipline.

Retarded children definitely above the border line of intelligence need thorough drilling in the techniques of reading and arithmetic more than they need activities. . . .

Academic handicaps on the part of normal but retarded children can be overcome by careful technical work on the part of teachers; but if such children are abandoned to manual arts and project activities they are thereby condemned to suffer throughout their academic career, and perhaps throughout life, from educational inferiorities which could easily have been overcome on the lower educational levels. . . .

"Learning by doing" is an excellent formula for inducing interest and effort in children and for awakening in them a consciousness of how the human race has materially progressed. It is the best method of learning any trade, profession, or art. *But has this formula any prominent place in the acquisition of the racial knowledge accumulated over immense periods of time, or in the development of abstract thinking?* . . .

Reading, rather than activity, is the way to erudition. . . .

The world had had "activity education" for six thousand historical years and knew very little at the end of that period. But during the relatively brief period when the world has been practising education by means of book-learning, its knowledge has grown apace. Humanity has learned a hundredfold more in the last three centuries than it had learned during the previous six thousand years.

Activity projects can be helpful in motivating our study and in preparing us to understand what we read. But nine-tenths—I would say ninety-nine hundredths—of what we moderns know comes to us from the printed page. . . .

Activity correlated with abstract thinking is the method par excellence of scientific discovery, in which observation and experimentation both inspire and verify ideas. We must grant that the educational functions of activity are valuable and indispensable. But *we cannot afford to let activity crowd out the functions of abstract education. Certain things can be learned much better through doing than thinking, but other things can be learned only through thinking.*[38]

Years ago Professor Dewey, in his *Democracy and Education*, also cautioned against a narrow interpretation of the activity idea:

Mere activity does not constitute experience. It is dispersive, centrifugal, dissipating. Experience as trying involves change, but change is meaningless transition unless it is consciously connected with the return wave of consequences which flow from it. . . . Blind and capricious impulses hurry us on heedlessly from one thing to another. So far as this happens, everything is writ in water. *There is none of that cumulative growth which makes an experience in any vital sense*

[38] *Ibid.*, pp. 115-120.

of that term. . . . The measure of the value of an experience lies in the perception of relationships or continuities to which it leads up. . . . There is no difference of opinion as to the theory of the matter. *All authorities agree that that discernment of relationships is the genuinely intellectual matter; hence, the educative matter.* . . . Thought or reflection, as we have already seen virtually if not explicitly, is the discernment of the relation between what we try to do and what happens in consequence. *No experience having a meaning is possible without some element of thought. . . . Thinking is the method of intelligent learning, of learning that employs and rewards mind. . . .*

Direct observation is naturally more vivid and vital. But it has its limitations; and *in any case it is a necessary part of education that one should acquire the ability to supplement the narrowness of his immediately personal experiences by utilizing the experiences of others. . . .*

Processes of instruction are unified in the degree in which they center in the production of good habits of thinking. While we may speak, without error, of the method of thought, *the important thing is that thinking is the method of an educative experience. The essentials of method are therefore identical with the essentials of reflection.*[39]

In their study of *child-centered schools,* Rugg and Shumaker likewise issue a word of warning against the avoidance of real work and of consecutive thinking, which is only too often the consequence of unrelated "activities." The following passages are of particular interest in this connection:

Emphatically, *intellectual development is avoided by many of these schools.* They stand for informality and they secure the outcomes of informality. *Their centers of interest* (they are well named) *lack intellectual rigor in plan and development. They are too often conspicuous examples of following the path of least resistance.*

Thinking is indeed hard work—there is no harder work—yet there is no royal road to understanding; prolonged intellectual effort offers the only route. The important meanings are difficult of comprehension. They will not teach themselves. The new school is obligated to teach them.

In the long run the intelligent person is the informed person. In the long run the intelligent person is he who has on tap a vast array of meanings, concepts, generalizations, and skills.[40]

The child as the center and the doctrine of freedom. Of even greater moment is the peculiar concept of initiative and child "purposing" which has come to characterize the radical activity schools.

[39] Dewey, John, *Democracy and Education,* Chaps. XI-XII. Excerpts are quoted by permission of The Macmillan Company, publishers. New York, 1917.

[40] Rugg, Harold, and Shumaker, Ann, *The Child-Centered School,* pp. 129 and 131, World Book Co., New York, 1928.

In such a school, the "child is the center." "Out of the child's own activity comes his growth. He needs contact with a rich environment, but the past can hold but little of value to him, for it must be secured in second-hand fashion, and second-hand experience can not be vital." Hence, "organized knowledge must be disregarded and the teacher must sit back and watch the activity go 'on apace.'"[41] In such a school, "the teacher does not direct." She has a "subordinate rôle."

In a later part of this study evidence will be submitted on how the doctrine of "child purposing," of "felt needs," and of "incidental learning" has affected our curricula and our standards. The displacement of the teacher's leadership by that of the child is, however, of such serious import that it merits further attention at this point. In a certain prominent activity school, so an official account states, the sponsors "never let a little thing like a program interfere with the day's business. . . . Everyone is intent on his own plans. The teacher is asked for advice, occasionally, but there is no unnecessary subservience to her idea."[42]

Common sense should detect the injustice of such a scheme both to the child and to society.

In the *first* place, *such a doctrine in its extreme form certainly will never work in a system of mass education.* A group of thirty or forty children without definite, collective guidance would soon degenerate into an incoherent mob.

Second, the activist's conception of freedom is not substantiated by modern biological science. The profound influence of the environment on a developing organism is now recognized.[43] It appears that it makes a tremendous difference whether or not a child's environment represents the type of milieu that he should have for his ideal development. This involves the idea of control and of guidance.

Third, it is far from clear how the momentary interests of any given child can be transformed into worth-while *consecutive* centers of interest for the *group,* or why one child should subordinate his own "felt needs" to those of his neighbor. In other words *how can social co-operation emerge in an atmosphere of unrestrained individ-*

[41] Cf. Hissong, C., *op. cit.,* [34], p. 12.

[42] *Ibid.,* p. 12. See also Burr, S. E., *An Introduction to Progressive Education,* Cincinnati, 1933.

[43] See Hissong's summarizing study of "Freedom in the Light of Modern Biological Foundations," *op. cit.,* [34], pp. 61-71.

ualism? Extremely serious problem cases are often due to the removal of uncompromising "law enforcement" in the home and in the schoolroom.[44]

Fourth, it is not true that all normal children crave unfettered freedom. Quite the opposite is the case, as is pointed out very effectively by Dr. Grace Adams. For seven years Dr. Adams worked as an assistant to a psychiatrist who specialized in "adjusting" the so-called "problem children" of the rich. In a revealing study she tells us that "children are at heart tremendous sticklers for things as they are and as they should be." Again, she discovered that the child "can be made to perform the most irksome task with the most eager pleasure if the performance is turned into a formal and complicated ritual."[45] Any experienced parent or teacher will readily endorse that statement. The golden-hued sentimentalism about the Child should be corrected, so Dr. Adams thinks, by a more realistic and truthful attitude. After stating that the average child may be expected to exhibit undesirable character traits, judged by adult standards, she says:

> The normal child is like this not because he has been poorly trained or badly conditioned or harshly repressed, but simply because he is a normal child. Most of us are inherently much more like him than we often care to confess; and we would resemble him even more closely if advancing years, and the experiences they brought with them, had not made us otherwise. But *progressive education, by shielding the child so vigorously during his early years, delays these chastening experiences until long after the more conventionally reared child has learned to adapt his nature to them. By this method infantile traits are certainly not repressed. But neither are they corrected. They are encouraged, made more pronounced, and prolonged* beyond the time when they should be gradually changing into characteristics more suitable to adult life.[46]

Basic Ideas of "Progressive" Education and Their Appraisal

The organization of the movement. The official sponsor of "progressive" education in this country is the Progressive Education Association. Organized in 1919, this association has grown tre-

[44] See Parsons, Alice Beal, "The Despotism of Polly Ross," *Harpers,* November, 1930, p. 669. Of significance, in this connection, is the modern play, *The Children's Hour,* which has been attracting large audiences for many months.

[45] See Adams, Grace, "Freedom and the Child," *Scribner's,* March, 1935, pp. 185 ff.

[46] *Ibid.,* p. 186.

mendously, now having a membership of approximately 7,000. Its official magazine, *Progressive Education,* started in 1924 as a quarterly but since 1929 has appeared as a monthly. The principal American supporters of the movement are such private schools as the Lincoln School of Teachers College, Columbia University, the Ethical Culture Schools, the Walden School, the City and Country Day School, all of New York City; the Tower Hill School of Wilmington, Delaware; the Beaver Country Day School of Chestnut Hill, Massachusetts; the Francis W. Parker School of Chicago, Illinois; the Fountain Valley School of Colorado Springs, Colorado. The Association is affiliated with the "New Education Fellowship," an international organization having representatives in at least twenty-eight countries, distributed over four continents. Many foreign periodicals are now devoted exclusively to the movement.

In the secondary field, progressive education is still distinctly "on trial," even more so than in the elementary field. There are indications that its original violence is already a thing of the past. Modified progressive education may prove to be a real help in making our secondary schools more truly cultural agencies than they are at present.[47]

Characteristics of progressive education. The principal tenets of the Progressive Education Association and of the New Education Fellowship may be summed up as follows:[48]

1. All indoctrination should be eliminated. Formal discipline and the learning of bookish facts should be abandoned. Vital activity must take the place of mechanical memorizing. *Learning by doing, by experiencing and experimenting, is of paramount importance.*

2. Since "only life can teach life," *the school must be the scene of real life, of physical activity, of making things instead of merely looking at them or hearing about them.* All barren intellectualism is to be eliminated.

3. *The gap between the school and life must disappear.* This requires the constant *introduction of life situations* and of new problems, largely discovered and selected by the pupil himself.

[47] For further details of the history of the movement, reference may be made to Burr, S. E., *op. cit.,* [42], pp. 7-9, and to the publications of the Progressive Education Association.

[48] Demiashkevich, M., *op. cit.,* [3], pp. 117 ff. For a more detailed statement of the principles of progressive education, see also the official publications of the Progressive Education Association.

4. *Only purposeful work will produce real learning and thinking.* Hence the pupil must make each problem his own and must work at it independently. All school work should be permeated by the spirit of play. *The customary techniques will then be acquired "incidentally."* The *what* of school work should be subordinated to the *how.*

5. The tyranny of established programs and of rigid time schedules must go.

6. *The child's immediate interests and desires should be held sacred.* They should drive out the usual "chalk and talk" of the traditional school.

7. The artificial stimulation represented by *marks, conventional report cards, rewards and punishments must be abandoned.*

8. *"Integrated" instruction is to replace the piecemeal approach by means of separate subjects.* The stream of life is a living unit which cannot be experienced and appreciated by its dissected fragments.

9. The teacher must cease being a "talking dictator." He is to aid and to listen rather than to command.

10. *Spontaneous activity on the part of the pupil must take the place of sequential work and organized curricula.*

A tentative evaluation. There are those who glorify these tendencies and proposals with all the enthusiasm of a religious faith. They are hopeful that these ideas will eventually transform our entire public system of education and not remain limited to a few highly favored laboratory schools. They look forward to a new educational paradise in which children will really be happy, in which there will be joyous activity instead of passive listening, and in which repression will cease. They see the arrival of schools that shall be the cradles of a new social order and of truly creative endeavor.

To what extent, then, is this program sound, and at which points is it utopian or dangerous in its fundamental assumptions?

It is obvious that, like the associated activity movement, the doctrines of progressive education rest primarily on Dewey's pragmatic instrumentalism. We have already criticized its doubtful theory of knowledge and "experience," its erroneous conception of "truth," and its rejection of first principles and binding standards or agreements. We have also pointed out the danger spots of the activity movement. All these weaknesses have been taken over, uncritically and almost blindly, by the confirmed followers of the "progressive" creed. The reaction that has already set in bids fair to prevent a major disaster to American education.[49]

Most certainly, we shall plan to retain the sound features and the corrective influences of the movement, without subscribing to its numerous flaws and eccentricities. Under conditions of mass education it is impossible to dispense with a considerable amount of discipline and control. When the community pays for the education of its children, it has the right to know what is going on in the schools. It has the right and the duty to guide its children in accordance with reasonable standards of personal and social efficiency. It cannot afford to give too much attention to private whims. It must make sure, first of all, of a wholesome, all-round product that will function successfully in later life. Is it not possible to harmonize such a more realistic view of education with the desire to see children happy, healthy, active, and even creative? Visits to the schools of the country, from coast to coast, would reveal that in thousands of classrooms this conception is already more than a dream. Progressive education has undoubtedly served as a pruning knife. It has made us acutely aware of our major educational shortcomings, our inflexible curricula, the perpetuation of meaningless drill, the questionable character of our testing programs, and the inadequate preparation of our teachers. To the extent to which the progressive education movement is correcting these drawbacks, it is placing us under a debt of lasting gratitude.

PART FOUR

THE CURRICULUM AND ITS BASIC FRAMES OF REFERENCE

A scene of confusion. A teacher entering on his duties for the first time discovers with some amazement, as soon as he has learned

[49] Attention should here be directed to the fifteen-year trial of progressive education in the Soviet schools. The plan was found wasteful, ineffective, and subversive of discipline. So glaring were its defects that it has been dropped in its entirety. For a description of this Russian experiment and the reasons for its abandonment, see *The New York Herald Tribune,* September 2, 1933; Bagley, William C., "The Task of Education in a Period of Rapid Social Change," *Educational Administration and Supervision,* November, 1933, pp. 561-578; Bagley, William C., "Is Subject-Matter Obsolete?" *Educational Administration and Supervision,* September, 1935, p. 409; Kandel, I. L., "The Educational Merry-Go-Round in Soviet Russia," *The Kadelpian Review,* May, 1935, pp. 327-334; Demiashkevich, M., *op. cit.,* [3], pp. 181-189.

to look behind the scenes, that there is no longer a feeling of confidence in the worth-whileness of his work. This feeling of uncertainty about the objectives of the school extends impartially to all phases of instruction. In fact, such is the instability of our educational convictions that we seem ready to scrap at a moment's notice plans that were heralded only yesterday as a glorious achievement. One day we become very much excited by the children's ignorance of the "fundamentals," and we straightway make copious provision for "remedial work." A month later, some one tells us that we are neglecting the creative talents of the pupils and their "individual interests and needs," and we hasten to make amends in that direction, often by curtailing the same "fundamentals" that seemed to require so much remedial attention. Our passion for educational novelties manifests itself as regularly as our habit of buying new spring hats or new winter coats. If we are not greeted each year at our educational mass meetings by some new setting, some readjustment, some "brand-new" crop of devices or slogans, we feel that something is radically wrong. At any cost, we must be "progressive." We have become "pragmatic" and "experimental" with a vengeance.

And so, "adjusting" the curriculum has become a major educational sport. "Everybody is doing it." Problems that have taxed the ingenuity of the wisest are boldly attacked and "solved" in the smallest hamlet. The latest curriculum-revision wave produced more than 35,000 courses of study, which are now on exhibition at Teachers College, Columbia University. And the end is not yet.

If this game did not involve the cultural and vocational welfare of millions of young people, it might be regarded as an amusing pastime, like playing checkers or pinochle. The situation looks immensely serious, however, when we become aware of its enormous cost, not only in dollars and cents, but in ruined careers and wasted years. We do not seem to realize that we cannot develop an integrated national life, a unified social consciousness, when the work of the schools is built on sand, on a fleeting impressionism, on momentary "felt needs," on local or provincial enthusiasms, on uninformed prejudices. When the curriculum becomes the football of powerful pressure groups, the result is chaos. It cannot be denied that many of our leading educators have been guilty of fostering this spirit of educational unrest, of change at any cost, even though they have been unable to offer more than temporary substitutes. In particular, at a safe distance from the classroom, and without first-hand experience, the curriculum "experts" have offered their wares with an assurance that is often inversely proportional to their actual competence. Real teachers have become weary of this costly and futile game, and they insist that there shall be less ballyhoo and more educational honesty, as well as genuine acquaintance with schoolroom possibilities.

Curriculum-Making without End

Curriculum revision and the American scene. The literature on curriculum-building in this country has grown to vast dimensions, and at an ever-accelerated pace. It has been asserted that this steady outpouring of suggestions for educational reconstruction represents merely a symptom of our national restlessness, our pioneering life in a new continent.

The intimate connection between curriculum-making and the "drama of American life" has been set forth with force and clearness by Professor Rugg. To quote:

> *Not once in a century and a half of national history has the curriculum of the school caught up with the dynamic content of American life.* Whether of colonial reading or reckoning school, Latin grammar school, academy, or modern junior high school, the curriculum has lagged behind the current civilization. Although the gap between the two has been markedly cut down in the last three-quarters of a century, nevertheless the American school has been essentially academic. To-day, much of the gap persists.
>
> *Not only has there been a huge gap between the curriculum and American life; a similar one has persisted to the present day between the growing child and the curriculum.* There are, indeed, three critical factors in the educative process: the child, contemporary American society, and, standing between them, the school curriculum.
>
> *Now, in more than a hundred years of systematization of the national educational scheme, the materials of instruction have not only been largely aloof from, indeed, foreign to, the institutions and culture of the American people; they have failed equally to provide for maximal child growth.* If the curriculum of our schools is to serve its true function, however, it must be reconstructed on a twofold basis. *Adequate provision must be made for creative personal development, and tolerant understanding of American life must be erected as the great guiding intellectual goal of education. Its reconstruction, therefore, must concentrate upon two foci—child growth and the dynamic content of American civilization.*

Now, from the early days of colonization, American life has been dynamic. With each succeeding generation the rhythm has accelerated.

> *The dominant theme is change, movement.* The innovation of today is relegated to the scrapheap of tomorrow. *Even national points of view are altered overnight. . . .*
>
> *The American tempo, I say, is prestissimo—and its intensity fortissimo. The current of American life is torrential.* It is personified by the pervading hum of motors and the dynamic syncopation of our new national music. *The American mind, like its industry, displays itself in movement, building, exploitation, "bigness and bedamnedness."*
>
> *In a hundred years, however, the public school has lagged far behind. It has never caught up with the momentum of industry, business, community life, or politics.* Only rarely has it succeeded in dealing with contemporary issues and conditions; *never has it anticipated social needs.* The masters of the American mind have fashioned the public school as a great conserving agency, and the halo of the past has oriented those who have made the content of our school curriculum. *Rarely have educational leaders affirmed for the school a preparatory and prophetic function.*[50]

So it appears that because of *our feverish national tempo,* the public school has forever "lagged behind." This outcry against our alleged backwardness is still with us. One recent example must suffice. In an address before the Progressive Education Association at its meeting in Hartford, Connecticut, May 12, 1934, Professor J. Ralph McGaughy, of Teachers College, said that "95 per cent of the schools in the United States are entirely out of step with contemporary life." Commenting on this statement, an able schoolman offered this reductio ad absurdum:

> If 95 per cent of the schools of today are out of step with contemporary life, then certainly that proportion must have been out of step ten, fifteen and twenty years ago—that is to say, 95 per cent of our citizens are out of step with modern trends. Who, then, creates the contemporary life? Surely, it is not the remaining 5 per cent. If anything, this statement proves just the opposite of what it intends—namely, that the emphasis placed in the public schools upon the basic principles of knowledge is most conducive to preparing the individual for contemporary life.[51]

To this apt reply may be added the query, how could the unquestioned achievements of American civilization have been accomplished by hopelessly inadequate schools?

Another perennial explanation of our supposed educational re-

[50] Rugg, Harold, "The School Curriculum and the Drama of American Life," in *The Twenty-sixth Yearbook of the National Society for the Study of Education,* pp. 3-4, Public School Publishing Co., Bloomington, Ill., 1926.

[51] Feingold, Gustave A., "The Basic Function of Secondary Education," *School and Society,* December 22, 1934.

tardation is that of *"college domination."* More recently, however, a well-known professor of education assured us that until twenty years ago we had been held in bondage by Europe, that we were just beginning to emerge from that intellectual slavery and were at last building a school system in harmony with the ideals of the fathers of our country. A strange alibi. Of course, the majority of our people, as well as the basic features of our civilization, originally came from Europe. It would have been decidedly queer if the pioneers had brought with them no educational ideas which they considered it desirable to perpetuate. But did that constitute "domination by Europe"? Do we not all belong to the great family of western nations that have a common cultural background? As the following pages will show, we have had ample opportunity and have made heroic attempts to achieve our own goals. What is obviously lacking, to this very moment, is clearness of purpose and a consistently applied educational philosophy. If we had a really convincing plan, our American initiative could be trusted to put it into effect.

Four decades of curriculum revision. The *Twenty-sixth Yearbook of the National Society for the Study of Education* gives a particularly illuminating survey of national curriculum-making during the past generation.[52] In this development we may distinguish at least three distinct types of procedure.

(1) *The Committee of Ten.* This committee (1892) was largely composed of college presidents and representatives of private schools. Of the ninety members of its subcommittees, forty-seven were college professors and administrators, twenty-one were headmasters of private schools, while only fourteen were principals of public high schools. The work of this committee exerted an enormous influence, but chiefly in *administrative* trends. Its emphasis was on college preparation, on uniformity and standardization, and on time allotment. The baneful effect of the unit idea, endorsed and imposed by the Committee of Ten, is still with us. Its *quantitative* formulation of an adequate college preparation has remained one of the chief stumbling blocks in the path of educational reform.

(2) *Era of National Committees.* This era extended approximately from 1895 to our own day, starting with the work of the Committee of Fifteen (1895) and continuing to the work of the Commission on the Social Studies, which is still in progress. Of

[52] See *op. cit.*, [50], Sec. I, Chaps. III-V.

particular interest to secondary schools were the contributions of the National Committee on Mathematical Requirements (1916-1923), of the Classical Investigation (1921-1925), and of the Modern Language Study (1924-1928). The *Twenty-sixth Yearbook* offers a penetrating critique of this phase of curriculum-revision. The shortcomings of the work done by these committees are attributed mainly to the fact that these committees "contained not a single professional student of curriculum-making, not an educational psychologist, not a sociologist, not a critical student of society."[53] This arraignment is then continued as follows:

> Recent national "subject" committees (like their predecessors) have not viewed the curriculum as a whole. Never have they taken a position aloof and tried to determine the vital *social* needs of children and adults. Instead, *they have been defenders of their particular faiths. They have been special pleaders for their subjects. They have maintained that their function was to inventory present practices.* The Classics Report put it: to investigate the "relevant facts of classics teaching" and to formulate programs for improving the teaching of their subject. They never really questioned the wisdom of teaching the existing content. They assumed that it should be taught. They stood like their predecessors of 1890-1920, for the status quo.
>
> *Never once did these committees open their minds to the really fundamental curriculum questions: Should mathematics and Latin be taught at all to all pupils? If so, on what grounds? To whom? With what materials? How chosen? On a broad analysis of the social needs of contemporary America or on the basis of disciplinary values?* Instead of answering these questions the reports (the Classics Report especially, the Mathematics Report much less conspicuously) used their vast arrays of facts to defend the position of mathematics and classics in the curriculum.[54]

As to the Mathematics Committee, we are told that "the personnel of the committee was destitute of professionally trained students of curriculum-making, hence the lack of objectivity and comprehensiveness in its procedures."

(3) *Nation-wide Curriculum-Making.* What the above comprehensiveness might mean was soon reflected in the *third* phase of national curriculum-making. It will probably go down in history as the great "curriculum jamboree."[55] Since the "specialists" had failed to produced acceptable curricula, the doctrine emanated from high places that perhaps the *classroom teachers* of the country

[53] *Ibid.*, p. 57.
[54] *Ibid.*, p. 59.
[55] See *Educational Administration and Supervision*, November, 1933, p. 568.

might find the answer to the great conundrum. Thus began an episode in curriculum-making which fostered the notion that "each community must have a curriculum all its own." The results were startling. As already stated, more than 35,000 different curricula have emanated from this movement, and new surprises are still announced from day to day. The idea of expecting classroom teachers without special training, without extensive library resources, and without an adequate time allowance, to fill the painful gap left by the specialists, was—in the frank works of Professor William C. Bagley—"not only silly but tragic." It is hardly necessary to enlarge on that pathetic theme.

The new approach. And now we are facing again the perennial challenge: What next? Well, the keen eye of Professor Rugg foresaw the only logical exit from the dilemma of curriculum-making as interpreted ten years ago. In Chapters III and IV of *The Twenty-sixth Yearbook of the National Society for the Study of Education*, he outlined his views of a really "scientific" procedure. Above all, we must have a "clear orientation as to the outcome of education":

> The day has passed in which a single individual professor, teacher or administrator, psychologist, educational law-giver or research specialist, can hope to master the manifold, highly professional tasks of curriculum-making. They are far too difficult and complex for any one person to hope to compass them all single-handed.[56]

There must be, according to Professor Rugg, two types of specialists who should take part in this co-operative enterprise: (1) Those trained in the study of the validity of materials; and (2) those trained in the science of curriculum-making, in the study of society and of educational psychology. He adds that "it is inconceivable that a curriculum can be made properly by either group working alone."

But who is worthy of taking part in this difficult task? Again, Professor Rugg supplies the answer, in the following passage:

> *The task of stating the goals of education is not to be consummated by an analysis of social activities alone. It will be aided by the latter, but must not be dominated by it. It will be achieved only by hard thinking and by the most prolonged consideration of facts by the deepest seers of human life.* For the great bulk of our curriculum, therefore, the analysis of social activities will influence the judgments of frontier thinkers; but it is the judgment of the seer based upon the scientific study of society—not the mere factual results of social

[56] Rugg, Harold, *op. cit.*, [50], p. 52.

analysis—that will determine the more intangible, but directing materials of our curriculum.

Social analysis merely gives us the techniques and knowledges we should have on tap. For the basic insights and attitudes we must rely, as we do for the statements of the goals of education, upon human judgment. It is imperative, however, that we make use of only the most valid judgments. The forecasting of trends of social movement, the perception of the focal problems and issues, and the connections underlying them, demand erudition and maturity of reflection that eventuates only from prolonged and scientific study of society. To the frontier of creative thought and of deepest feeling we go for guidance as to what to teach.[57]

It would seem, then, that we are now ready for a *fourth* stage of national curriculum-building, that of the "frontier thinkers." Evidently the challenge has been heard. The mantle of wisdom has obviously descended on a group of self-appointed disciples who feel the urge to lead us out of the wilderness into a new land of promise. Unfortunately, the new symphony of expert talents seems to be marred by occasional dissonances, perhaps due to faulty orchestration. At any rate, a unified melody has not yet emerged. There are, in fact, quite a few versions of the new theme song.

But the main point is that the central secret has already leaked out. The solution now proposed is as amazing as it is simple. It is to the effect that, since the curriculum has caused so much trouble, we can improve the educational situation only by *abolishing the curriculum.* This may sound like magic to a poor, unsophisticated school teacher. It is like curing the dog by the famous expedient of cutting off his tail just behind his ears. The arrival of "Nirvana" may, indeed, be the final attempt at curriculum tinkering, for there will be no "subjects" left for further benevolent execution. And so the curtain is about to rise on another New Deal in education, the *planless school.*

New Ways of Dealing with the Curriculum

The planless school. The arguments which underlie this latest development represent a maze of partial truths and gross errors. Here we meet again our old friends, already considered at length, namely the doctrines of pragmatic instrumentalism and the programs of the activists and the social reconstructionists.

Immediate experience alone is considered worth while. Momen-

[57] *Ibid.*, p. 82.

tary interests and problems transcend in importance the experience of the race. Individual freedom and "felt needs" must be respected at any cost. Standards of excellence are necessarily outmoded. Sequential learning is of no consequence and is regarded as stultifying. The traditional "subjects" have been kept in the schools, we are told, only because of a blind belief in the "obsolete" doctrine of mental discipline. We must drop that hoary myth and with it throw out foreign languages, mathematics, and other "disciplinary" subjects. Everywhere and at all times the emphasis is to be on *social values, social efficiency and participation.*

When the school has been thus emancipated from its traditional fetters, progress can at last be made in achieving the *real* objectives of education. It is true that these have not yet been fully formulated, but they are expected to emerge in due time, after further experimentation and "reconstruction." The new school will be very different from anything we have had thus far. Glimpses of what is intended may be found by those who are curious by reading some recently published monographs dealing with that subject.[58] The aim of the transformed secondary school is described by one of its enthusiastic proponents in the following manner:

> The new high school should have as its central and dynamic objective: *to prepare young people to take an active part in planning and building a new society which fulfills the material and cultural needs of the great mass of the American people.* This objective is definite, meaningful, and dynamic, a guide and a drive to action.[59]

As to a program, essentially that of "social reconstruction," we are informed as follows:

> In planning the community and the school there should be a long-term plan which states in broad outline what the needs of the people are. There should also be a short-term plan extending say over sev-

[58] See, especially, Everett, S. (Editor), *A Challenge to Secondary Education* [22].

[59] *Ibid.*, p. 198. For further information concerning this "social" objective of education and related problems, the reader may be referred to such sources as the following: Kilpatrick, William H. (Editor), *The Educational Frontier,* [20]; *The Social Frontier, a Journal of Educational Criticism and Reconstruction,* especially the issue of January, 1935; Childs, John L., *Education and the Philosophy of Experimentalism,* The Century Co., New York, 1931; Overstreet, H. H., *We Move in New Directions,* W. W. Norton & Co., Inc., New York, 1933; Reisner, Edward H., "Can the Schools Change the Social Order?" *Teachers College Record,* February, 1935; Counts, George, *Dare the School Build a New Social Order?,* The John Day Co., Inc., New York, 1932, representing No. 11 of The John Day Pamphlets; and, especially, *The Thirteenth Yearbook* (1935) *of the Department of Superintendence,* "Social Change and Education," Washington.

eral years which is followed by succeeding short-term plans. Step by step each plan leads toward the goals set in the long-term plan, which of course is revised in the light of the experiences accumulated in working out the intermediate plans.[60]

The curriculum is conceived of as a series of "experiences and activities":

> Instead of organizing the curriculum into narrow subjects of study, it should be set up as integral parts of the broad theme or objective, namely, "*planning and building the new society*." For convenience of reference, these parts might be designated as areas of experience, for example, "making a living," "developing health," etc. Each of these areas would have definite social objectives which would be approached through a consideration of crucial life problems and through participation in vital activities.[61]

The many proposed "activities" of the new school are justified in the following manner:

> Two fundamental principles of modern psychology are that learning is an active process and that people learn what they practice. In accordance with these principles, the new high school should have an abundance and variety of student activities.[62]

As to the *equipment* of the new school, we are told that it will be converted from a "factory" into a place for "living." There will be movable tables and chairs. Each classroom will have its library, its newspapers and magazines. Special rooms will be available for particular "areas of experience." There will be other rooms where pupils may write poems, draw or paint, or "just rest." Of course, radios will be available at every turn.

In such a school there will be "no hard and fast grades" and there will be little or no emphasis on marks. The pupils themselves are to be judges of their own achievement. Above all, they will spend considerable time in making first-hand studies of the community. They will visit factories and offices, slums, museums, and city halls. They will learn, as directly as possible, all about family and community budgets. Their knowledge of economics, government, and social planning is to become real and "functional."

And what about the *teacher* in such an emancipated school? It is recognized that the new program requires "master teachers." Their main function would be that of *guidance*. Each master

[60] Everett, S. (Editor), *op. cit.*, [22], p. 200.

[61] *Ibid.*, p. 200.

[62] *Ibid.*, p. 208.

teacher would be responsible for the work of approximately a hundred pupils. He would be a "teaching counselor, remaining with his group of pupils throughout three, four, or six years."[63] There should also be *ordinary* teachers who are to "advise a group of fifteen or twenty students." "The adviser is responsible for seeing that pupils really work; that they put themselves wholeheartedly into projects; that they accept and carry out duties and responsibilities. He stands in lieu of requirements and administrative pressures." Quite a contract for the "ordinary" teacher! In fact, an "imaginary typical day" for a teacher whose special field is *"statistical mathematics"* is suggested in Professor Goodwin Watson's monograph.[64] The schedule extends from 9 A.M. to 8 P.M., when at last the harassed teacher-adviser is "free for recreation *and study*."

Each pupil has a card file of desirable projects, under various "functional" headings. From these he will select certain ones *"whenever an idea appeals to him."* Some of them represent "long-term purposes," and some "short-term purposes." Professor Watson submits a tentative list of seventeen "functional divisions" of the curriculum. They include such diverse items as "establishing a home," "bringing up children," "working out a satisfying philosophy of life."[65] Evidently, mere subject matter and ordinary textbooks will not be necessary or sufficient in developing such "functional" areas. Each will require "several hundred guides," and "new guides will be added constantly."[66]

To get this whole enterprise started smoothly, Dr. Samuel Everett wisely remarks that "a social philosophy must first be formulated and accepted." As to separate subjects, however, he says with finality:

> Most Americans will *never* use history, mathematics, physics, chemistry, and the like, as these subjects are now taught. For the most part *such skills and knowledge can be learned in a graduate school*, or *whenever individuals feel the real need for such specialization.*[67]

This is another return to the theory of "incidental learning," disproved so often. The same writer achieves even greater heights, or depths, of insight, when he writes:

> Few Americans will ever travel abroad and therefore need foreign language. A relatively few individuals will use foreign language in

[63] *Ibid.*, p. 63.
[64] *Ibid.*, pp. 162-163.
[65] *Ibid.*, p. 154.
[66] *Ibid.*, p. 159.
[67] *Ibid.*, p. 295.

business. Only a very small group of children will love language for its own sake. The whole tone of a secondary school can be so developed that *the acquisition of foreign language as a mark of the leisure class and as a badge of invidious distinction will be in disrepute.*[68]

One wonders whether this gentleman ever saw one of the huge ocean liners that annually carry a million American tourists across the sea. One wonders whether he has ever witnessed an assembly program given by modern language clubs in our high schools, organized by pupils who evidently enjoy the study of foreign languages. And one doubts whether this astute observer of the American scene realizes that many thousands of young men and women in our professional schools, as well as all research workers and technicians in countless laboratories, must, if they are to be leaders, make wide use of foreign books and periodicals. Similarly in mathematics. Is such training also to be acquired "incidentally," as a sort of adult "project"?

We have given enough samples of what is intended in the "ideal" secondary school of the future. After learning of these plans, a clever teacher remarked: "These educators are shining examples of the *minimum education* which they wish to force on our American schools, the only type of education which they seem able to comprehend." When the whole world is clamoring for more complete, systematic, and thorough training in every field of human activity, we are advised by these educational spokesmen to develop a planless school devoted to "functional" areas adjusted to "personal interests" or alleged individual "needs."

Curricula based on orientation. Much more significant is the emphatic plea for educational or professional orientation, which is encountered again and again in the recent literature on secondary education. Many subject-matter teachers have, of course, been keenly aware of this legitimate demand. They know that each subject is, or should be, a constituent and vital part of the great symphony of personal and social interests. Only too often a fragmentary and piecemeal mode of teaching obscures this all-important fact. The spectacular success of such books as Wells's *Outline of History* demonstrates the positive hunger for a panoramic view of things on the part of countless thousands. Apparently, this age of the radio, of air travel, and of the movies has prepared us for larger visions and more inclusive discussions. We all know of the

[68] *Ibid.*, pp. 295-296.

danger of extreme specialization, so characteristic of our age. The specialist only too often is a person who "learns more and more about less and less." The story is told of a bone specialist who was called to see the victim of an accident. The poor fellow had broken his right leg. Shrugging his shoulders, the specialist refused to take the case because he was merely an expert in "left legs."

Besides, the fund of human knowledge has become so vast that no single human being can cope with it alone, even in restricted fields. Thus, Dr. Alexis Carrel tells us that a comprehensive mastery of the essentials of the new "science of man," as he conceives it, would call for the unremitting, painstaking labor of at least twenty-five years on the part of even a super-scientist.[69] Similarly, Dr. E. T. Bell informs us that modern mathematical science has grown to such dimensions that "a detailed, professional mastery of the whole domain would demand the lifelong toil of twenty or more richly gifted men."[70]

For some time, the colleges have recognized the imperative call for orientation. More than four hundred colleges are now giving survey courses in one form or another. But how can the *secondary* school meet this demand? How can one "orient" pupils who have not yet acquired even the rudiments of a general education? At least two plans, differing widely in scope and in merit, have emerged. The first is a debatable imitation of the orientation courses offered in the colleges. When high schools offer "exploratory" or "survey" courses extending over a few weeks or months, they are in great danger of encouraging a superficial smattering of subject matter that is imperfectly understood and quickly forgotten. Such courses may be entertaining and even "popular." They can never be a substitute for more substantial, long-continued contact with any one of the major subjects of the curriculum. Above all, they undermine the appetite for really serious study by creating a false sense of mastery. They are reminders of the brief "exposure" courses given in the academies and high schools several decades ago. Such texts as "Fourteen Weeks in Botany" rarely left a discoverable trace in the pupil's mind. Why revive a plan that was discarded long ago as futile? And yet, it is now proposed that "slow pupils," especially, be given precisely that sort of "snappy

[69] Carrel, A., *op. cit.*, [27], p. 285.
[70] Bell, E. T., *op. cit.*, [25], p. 5.

orientation." In mathematics, we already have a number of curricula and texts based on this idea. Primarily for "slow" pupils or "non-academic" pupils, a one-year survey course in secondary mathematics is now proposed, involving the elements of business arithmetic, "applied" algebra, intuitive geometry, trigonometry, demonstrative geometry, statistics, and copious related units of an optional type. In some cases, morsels of physical science are also to be included. Any teacher of mathematics knows that such a one-year program is dishonest and impossible. It will lead only to the further destruction or disintegration of secondary mathematics.

Of very different quality is the type of orientation that gives the student a real understanding of a major field in which he may be interested. It may be described as "intensive orientation." Such a plan has been developed at the Fieldston School, New York City. After seventeen years of experimentation, the plan is now working successfully. In a thirty-page monograph, Dr. V. T. Thayer, director of education of the Ethical Culture Schools, has given us a welcome description of this program.[71] Thus, in the field of art, a student may pursue a four-year, preprofessional orientation course. Any major center of interest of this type is then used as "a bridge to culture,"[72] leading to many extensive contacts with science, history, mathematics, and other related branches. A student thus prepared carries away not only a substantial equipment in his field of concentration, but also in many of the other basic domains of knowledge and skill. It is the type of orientation that one might welcome in any high school.

The core-curriculum plan. Closely resembling the program of intensive orientation is that advocated by proponents of the *core-curriculum* idea. But while the former is essentially prevocational in character, the latter is very general and is intended for every student. Various modifications of this plan have appeared in print. One outstanding example, the result of the Virginia Curriculum Program, is described by Mr. Sidney B. Hall, Superintendent of Public Instruction in Virginia, in collaboration with Principal F. M. Alexander.[73] For all pupils in the secondary school a core-curriculum has been set up which is organized around certain broad

[71] Everett, S. (Editor), *op. cit.*, [22], Chap. IV.
[72] *Ibid.*, p. 94.
[73] *Ibid.*, Chap. II.

fields of knowledge. It is based on centers of interest that are definitely related to eleven major functions of social life. The curriculum extends throughout all the years of the high school period. Definite provision is made for those pupils who also wish to "elect the study of specialized organized bodies of knowledge such as foreign languages, mathematics, and so on." Various "adaptations" of the plan are suggested. It is noteworthy and very encouraging that vocational education is definitely ruled out of the picture. To quote:

1. *The scope of the curriculum for the entire system of public schools should be organized as a continuous system on the basis of the major function and center-of-interest approach.* The secondary school should be merged into this common-school program and should not be regarded as a separate institution with peculiar functions of its own.

2. *Vocational education except commercial training, other types of training on this level, and exploratory work, should be postponed until the period of general education is over. This means that general education would normally continue to what is now considered the end of the junior-college period.*

3. Pupils should be retained until in the judgment of school authorities, unguided by the school, they can discharge responsibilities to society.

4. This common-school program should be concerned with the value of its own courses. *The courses should not be aimed toward preparation for vocational work or advanced studies.*[74]

A divergent and more conservative plan is that suggested by Dr. W. B. Featherstone, formerly Director of Secondary Curriculum, Los Angeles, California, and now Assistant Professor of Education at Columbia University.[75] It is definitely recognized that there are common educational essentials which all persons must have acquired in the interest of social solidarity and the security of life itself. More than that, the importance of *continuity* in the major activities is admitted. Beyond that, each pupil may pursue such "marginal" activities as may appeal to him particularly. The one great enemy of such a core-curriculum is seen to be the current *unit system.* In the following paragraph our present dilemma is clearly pointed out:

There are many elements which belong in the core-curriculum—English, history, art, science, music, geography, economics, psychology,

[74] *Ibid.,* p. 47.
[75] *Ibid.,* p. 67.

and probably others. Obviously we cannot provide a period a day for each of these and have any time left for other matters. Furthermore, we dare not spread them out, providing a dose of one here and a dab of another there. If we do, we shall violate one of the most important principles of teaching, that of *continuity*. There is only one alternative, and that is to develop unified programs which bring into one learning situation essential materials that were formerly found only in subject-matter fields sharply separated from one another. We must find a thread in the form of *a central theme or central idea* upon which we can string the separate beads of language, geography, science, and the like, and have when we have finished something that is whole in its own right and not just an aggregation of unrelated entities.[76]

The integration movement. A further variation of the same general idea is represented by recent insistent demands for a greater *integration* of our instructional enterprises. In this case we may have in mind a core-curriculum such as those described above, or we may think of the fusion of two or more otherwise distinct lines of work. The latter plan has been advocated particularly by educators connected with Teachers College, Columbia University. Thus, Professor L. Thomas Hopkins has submitted a detailed list of the characteristics of an "integrated individual" and of the curriculum changes which this concept seems to imply.[77] At present, one of the most active supporters of this plan is Dr. Jesse H. Newlon of the Lincoln School of Teachers College. The following typical passages from one of his recent addresses on this subject, as quoted in *Teachers College Record*, may be of interest:

It is of critical importance at this particular time that the curriculum be designed to make possible the achievement of the broad social values of education.

To this end an effective plan must comprehend all significant cultural processes and problems in order to give the individual the best possible orientation to the world in which he lives. It must introduce him to all important areas of human knowledge and experience. There must be no gaps in his social education. . . .

Integration of the experiences and activities of the learner at each level is also essential. These are more than problems of grade placement of materials. They are not problems in mechanics.[78]

Experiments in this direction have been going on at the Lincoln School since 1929. Attempts have been made to integrate mathe-

[76] *Ibid.*, pp. 67-68.

[77] See Hopkins, L. Thomas, "Arguments Favoring Integration," *Teachers College Record*, April, 1935, pp. 604-612.

[78] *Teachers College Record*, March, 1935, pp. 518-519.

matics and science in the seventh grade. In 1932 a course devoted to a study of the evolution of Western culture was introduced, replacing the usual courses in history, English, and art. This course is expected to culminate in an intensive study of American culture. Needless to say, all the usual subject-matter boundaries are deliberately ignored in these experiments.

Comments on these plans. First, as to the planless school, it is to be hoped that our proverbial sense of humor may keep us from taking this educational burlesque too seriously. However, if heavier artillery is needed to demolish this almost incredible extravaganza, there is an ample stock of it for such bombardments. For example, let the reader turn to such devastating critiques as those furnished by Professor Bagley, Professor Kandel, and Professor Demiashkevich.[79] Above all, let him learn of the verdict of Russia when, after nearly fifteen years of a compulsory, nationwide trial of planless education based on the pattern glorified by our own theorists, it swept the whole useless structure out of existence.[80]

We may, however, feel much more optimistic about the other three plans discussed above. In due time, their extreme and visionary features will be eliminated. Orientation of the Fieldston type has great possibilities. The core-curriculum idea has the merit of not giving up an emphasis on essentials. And integration has always been a more or less subconscious objective of every real teacher. The "many-sided" interest of the Herbartians and subsequent "correlation" theories have served to keep this ideal alive in modern education.

On the other hand, any attempt at excessive *fusion* will almost certainly lead to failure. There have been numerous previous attempts in that direction. They have never been successful. Thus, the plan of fusing mathematics and science was launched with much enthusiasm, thirty years ago, by a group of leaders in the Middle West. A special committee was organized which eventually submitted its recommendations in a published report.[81] Much lab-

[79] See Bagley, William C., "Is Subject-Matter Obsolete?" *op. cit.* [49], pp. 401-412; Kandel, I. L., "Alice in Cloud-Cuckoo-Land," *Teachers College Record*, May, 1933, pp. 627-634; and Demiashkevich, M., *op. cit.*, [3], pp. 245 ff.

[80] See references in footnote 49, page 58.

[81] Report of the Committee on the Correlation of Mathematics and Physics in Secondary Schools, Second Edition, 1905. This constitutes the *Appendix of the Proceedings of the Second Annual Meeting of the Central Association of Science and Mathematics Teachers* (out of print).

oratory work was suggested and particular endorsement was given to the famous Perry movement, then at its height in England. Professor Perry's main objective was in the direction of applied mathematics in a thoroughly correlated setting. Hardly a trace of that movement is left in England, in spite of the enthusiasm with which it was greeted by many teachers. Must we always repeat the errors of others?

A critique of integration, in its relation to mathematics, was recently offered by Professor W. D. Reeve.[82] He furnished a helpful bibliography and pointed out some of the merits and dangers of the movement. Of particular importance is the following passage:

> It should be said here that before teachers can properly correlate mathematics with other fields, they ought to learn how to correlate the various parts of mathematics. They should first learn how and where arithmetic and informal geometry can be correlated, how and where algebra may be best correlated with arithmetic and informational geometry, and so on. Unless we can do this, there is small chance that we can successfully correlate mathematics with science, music, the arts, and other applied fields. We should not make the mistake of breaking down subject-matter lines before teachers are qualified to make the proper contacts and to explain the significant relationships between subjects to their pupils.[83]

Teachers cannot integrate several subjects or courses unless they have mastered each of them. At present, it is only too clear that we have not yet developed adequate mastery in even one field on the part of the majority of our teachers. Are they to spend three more years in graduate schools for the sole purpose of integration? What salary inducements, and what professional security are we offering in return? *It is significant that in most European countries teachers of science are also the teachers of mathematics. In spite of this ideal preparation, they prefer not to fuse these subjects.* They realize that each of these fields has its own genius and its own techniques. Science and mathematics should constantly support each other, but they would only lose their characteristic educational functions by being fused into a single subject or "area of experience."

Basic Frames of Reference

Unchanging backgrounds. After days or weeks of weary travel through an arid desert, many a caravan has been spurred on

[82] See Reeve, W. D., "Mathematics and the Integrated Program in Education," *Teachers College Record,* March, 1935, pp. 497 ff.

[83] *Ibid.,* p. 503.

by the realization that an oasis promising fresh water and a source of food was not far distant. Just so, in our moods of despair over the educational situation, we may gain renewed courage and vigor by keeping our eyes on those enduring landmarks that never fade from the picture. For there are three such mountain peaks. They are Nature, Man, and the Ultimate.

In the realm of *Nature,* we witness everywhere the reign of dependable laws. The cosmic force of gravitation prevailed before the arrival of man, and we may count on its further operation in all the aeons that may lie ahead. The sun and its satellites have been spinning through space for untold millions of years. Life-giving and life-sustaining principles and energies have been at work during countless ages, and a belief in their beneficent continuance is not at variance with accepted postulates of science.

As to *Man,* the mysterious stranger on this strange earth, we know at least that his physical needs, his dependence on food, clothing, shelter, and other material comforts, as well as his social needs, will always remain the foundation of his ceaseless struggle to triumph over hunger, want, limitation, and disease. Long and painful has been his pilgrimage, terribly exacting has been the tuition fee for his education, many are the lessons still to be learned. But ever more clearly there emerges the vision of a social humanity created by truly wise, self-controlled, co-operative men and women.

Finally, we have been on the road long enough to feel increasingly sure of the permanent presence of a *third* mysterious element which is "nearer than breathing, and closer than hands or feet." "Some call it evolution, and others call it God." However we may wish to paraphrase it, whether we call it the Supreme Mind, or the Infinite, or the Good Life, the deepest seers of mankind have reverently bowed to its actuality and have attributed to it their own inspiration. It is here that we encounter a third permanent realm, the universe of ideals and moral values, of high endeavor and unselfish devotion.

The one great variable. What, then, we must now ask, is it that changes? The answer is simple. Man's *understanding* of his immense environment changes. His growing *insight* into the physical, social, and spiritual *relationships* that condition human welfare has been the great transformer that continues to lift him from savagery to ever greater heights. It has enabled him to overcome his serfdom to the forces of nature, his bondage to fear and superstition, and to develop an ever more confident control of his destiny. *Human intelligence is not a constant. It is the great variable.* Education is forever concerned with the development and the transmission of higher and nobler forms of insight into the basic relationships mentioned above.

Constants of the curriculum. If this analysis is sound, it follows at once that the curriculum should be the vehicle of a broad orientation. It should initiate in each individual a growing ability to understand, to appreciate, and to control his environment, not for purely selfish reasons, but always in the direction of an increasing social welfare. Hence the permanent elements of the curriculum must be those manual, mental, and spiritual adaptations which race experience has shown to be essential for human progress and for an integrated social life. *We must have at least six such elements in an elementary curriculum, as follows: the practical arts, language, science, mathematics, the social studies, and the fine arts.*

1. *The practical arts.* All the basic achievements of our material civilization are fundamentally products of the human *hand.* The machine is nothing but an extension of the hand. The human hand and its substitute, the machine, are forever concerned with the utilization and the transformation of the raw materials furnished by nature, for the immediate and ultimate purposes of society. The epic of the hand still remains to be written. The story of the transition from the crude implements of the Stone Age to the marvels of a modern machine shop is an epitome of human suffering and human triumphs. It is one of the tragedies of our big cities that many children and adults no longer experience the dignity and the satisfaction of real handwork. We must restore some of the lost opportunities for obtaining such training. Not for vocational purposes, but for the keen enjoyment and appreciation of the ability to "make things" with one's hands, and not merely with a machine, we must furnish in each school a reasonable variety of creative *manual* activities. The Universal School of Handicrafts, in Rockefeller Center, New York City, directed by Edward T. Hall, might well serve as a model for a similar organization in both large and small high schools.

2. *Language.* The spoken, written, and printed word was the magic wand that made possible all forms of *symbolic* thinking. It transformed a crude herd instinct into the miracle of social com-

munication. Dewey has stated correctly that *the invention of symbols was the greatest single event in human history.*[84] The preservation of human accomplishments by means of written or printed records made man a "time-binding animal," to use a famous phrase due to Count Korzybski.[85] Hence we have good reason to stress the transcendent importance of the language arts in the curriculum. Here we may also include the study of foreign languages, if time permits, for—as a great thinker once said—whenever we study another language, we add another soul to our own.

3. *Science.* We have already pointed out that science is concerned with *facts.* The aim of the physical sciences is the discovery and formulation of the relationships which govern the happenings of nature. Hence science is the handmaid of technology and of all forms of engineering. When applied to the "human machine," it leads to the life-conserving study of health. The biological sciences, including psychology, have opened up a new universe that the school is only beginning to appreciate. Modern life depends on science at almost every step, and the curriculum must make adequate provision for broadly conceived contacts with scientific backgrounds, including a constant emphasis on bodily health and efficiency.

4. *Mathematics.* In co-operation with science, mathematics has furnished the principal tools for discovering, testing, and stating the laws of nature. It is *the abstract form of science, the prototype of all science.* Without it, our material civilization would crumble into dust, and man would return to a state of savagery. These undisputed facts alone would warrant a mandatory emphasis on mathematical training in our schools.

5. *The social studies.* It has been charged that in the unavoidable endeavor to understand and control our *physical* environment, we have neglected the study of man as a *social* being. To this alleged fact leading educational critics now attribute many of our present-day difficulties. As a corrective they insist on a much larger, an almost exclusive, consideration of the *social studies* in the curricula of our schools. While we must reject extreme demands in this direction, it is certainly true that the citizens of a democracy surrounded by dangerous world currents should be informed as to the fundamental social, economic, and political problems of the

[84] Dewey, John, *The Quest for Certainty,* [6], p. 151.

[85] Korzybski, Alfred, *Manhood of Humanity,* E. P. Dutton & Co., New York, 1921.

immediate future. For a time, at least, this social emphasis may have to replace or supplement the customary pageant idea of human history, the orderly recital of man's evolution from a primitive state to that represented by modern civilization.

6. *The fine arts.* Music and the representative arts, aside from literature, constitute a form of release from life's daily routine. Here, at least, we have a field of almost pure creativeness and of ideal appreciation. The liberating influence of the fine arts can hardly be exaggerated. No human being should be sent out into life without having passed through the door that leads to this land of enchantment.[86]

These six domains constitute the permanent frame of reference for all curriculum-making. *They are basic because they have grown out of the very nature of things.* They are the products of human evolution. *They all date back to the beginnings of human history. And they will continue until the end of the story.* If all human institutions were destroyed today, provided human beings remained, these six great basic categories would return with inevitable certainty. Not one of them, therefore, can be omitted from the curriculum, for that would cause the loss of an essential "area of experience."

We have not included in this list the all-important matter of *character building*, simply because ethical education does not constitute a "subject" or a formulated course of study that can be mastered from day to day by a stated series of lessons. The school must certainly make a definite contribution in this direction.[87] But it can never relieve the home and other agencies from their share in developing strong, moral characters, from stressing ethical and religious ideals, and from discouraging all forms of antisocial or criminal behavior.

The time element. Ever since there have been modern schools, there has been the puzzling question of how the time could be budgeted to accommodate all the desired objectives. In the modern

[86] The remarkable success of broadly constructed and well-organized art courses in general or specialized high schools is a welcome proof of the fact that many young students elect these studies, not because they are supposed to be "easy," but because they offer an essential and enjoyable element of cultural training. Among the schools which are doing pioneer work in this direction is the Newark (New Jersey) Arts High School.

[87] See "Character Education," *The Tenth Yearbook of the Department of Superintendence,* Department of Superintendence of the National Association of the United States, Washington, 1932.

American high school, *the solution has been made virtually impossible by the unit system which is still in force. It must be given up. Instead, we must have continuity of work in each of the six trunk-line domains outlined above, over a period of at least four years, from grade seven through grade ten, and preferably through grade twelve.* To accomplish this, we must assign three or four periods a week to some of the major subjects, instead of five. This will leave a flexible margin of time for health training, and for work in the practical and fine arts. Schools preferring a core-curriculum, or orientation courses in minor fields, will then be able to work out useful variations and still provide for continuity in the major fields.

If it be objected that this cannot be done, attention may be directed to the time-schedules of other leading countries, as published in such sources as the *Educational Yearbooks of the International Institute of Teachers College*, the *Report of the National Committee on Mathematical Requirements*, and the *Fourth Yearbook of the National Council of Teachers of Mathematics* (1929). All these "time-tables" provide for continuity in the principal subjects, and especially in language, mathematics, and science. The total number of prescribed weekly periods usually exceeds that of the corresponding American schools. For example, in the Japanese secondary schools (middle schools), the boys carry thirty periods of work each week, and the girls have twenty-eight periods of weekly instruction. Again, the number of prescribed periods assigned to mathematics in schools for boys, each week, *throughout a period of five years,* is either four or five.

The Unique Importance of Mathematics

The crucial question. We have referred previously to the growing tendency to throttle a defense of secondary mathematics by arguments that are anything but conclusive. Of course, well-informed educators should realize that in no other country is such a defense of mathematics considered necessary. The earliest and most far-reaching of all the sciences needs no "special pleading." It will continue to be a core-subject long after the vagaries of our day have disappeared. However, many of our boys and girls are now being deprived of an essential element in their education by a misdirected propaganda against mathematics. And that is a very serious matter. Seventeen states have already eliminated mathematics from the list of prescribed subjects in their secondary schools. Among the principal reasons given for this action by responsible officials are the following: (1) poor teaching, (2) poor textbooks, (3) a lack of essential and convincing objectives.[88]

Now there is hardly any doubt that the accusation of "poor teaching" may be extended with equal justice to any and every other secondary subject. Perhaps the same thing is true also of the textbooks. So it comes down to this, that our courses of study evidently do not seem to reflect convincingly the real spirit and the purpose of mathematical instruction at the secondary level. To put it differently, the question that we are now facing is the following:

Precisely what is the real or potential contribution of mathematics that would justify its mandatory inclusion in the education of all our young people, even if it be conceded that many of them may never have occasion to use mathematics for strictly vocational purposes?

The service value of mathematics. The answer may be resolved into two parts. We must first look into the *"instrumental"* values of mathematics, for that is the only phase of the subject that seems to interest the man in the street and the pragmatic type of educator.

Well, that part of the story has been treated so many times by scores of competent writers that any one who can read may inform himself along this line. Merely ignoring this evidence is not a refutation of its truth.

In the Century of Progress Exposition at Chicago there was exhibited on one of the walls of the Hall of Science a large picture called the "Tree of Knowledge." Its roots and branches represented the "Basic and Applied Sciences." At the foot of the tree, forming its very foundation and main root, was placed the legend, "Mathematics." This picture may now be seen at the new Museum of Science and Industry in Chicago.

On the wall opposite that picture, the observer was confronted with the story of electricity and of the radio. This account was written by officers of the United States Navy and Marine Corps. It culminated in the following sentence: *"Mathematics is the Key and Applied Mathematics is the Tool wherewith Man conquers the Universe."*

[88] See reports on the present situation in *The National Mathematics Magazine*, October and November, 1934, and February, 1935.

This same theme is treated in almost endless variation in scores of articles or reports found in such publications as *The Mathematics Teacher, School Science and Mathematics, Scripta Mathematics,* and the *Yearbook of the National Council of Teachers of Mathematics.*

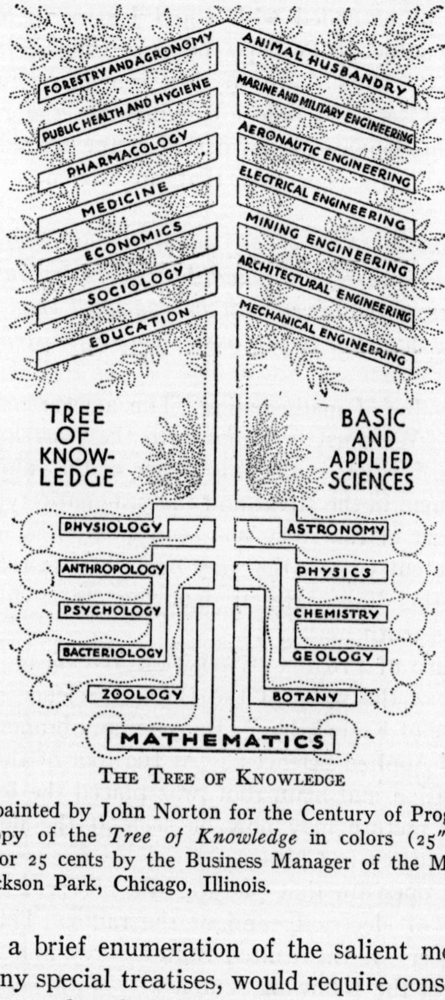

THE TREE OF KNOWLEDGE

From a mural painted by John Norton for the Century of Progress Exposition

An enlarged copy of the *Tree of Knowledge* in colors (25″ × 38″) will be mailed postpaid for 25 cents by the Business Manager of the Museum of Science and Industry, Jackson Park, Chicago, Illinois.

Even a brief enumeration of the salient monographs, not to mention many special treatises, would require considerable space.

It may, however, be of greater interest to hear how some of our higher institutions of learning are reacting to the current attacks on mathematics. Thus the University of Wisconsin recently took official action concerning the present situation by a drastic modification of its methods of admission. Hereafter, students are admitted on either an "unrestricted" or a "restricted" basis, according as they have, or have *not*, met the usual requirements in mathematics. Those who come without such preparation, may pursue such studies "as do not require high school mathematics as [their] background. It does *not* give admission to the College of Agriculture or the College of Engineering or the Course in Chemistry, and does *not* permit the student to major or specialize in chemistry, commerce, economics, mathematics, pharmacy, political science, pre-medicine, philosophy, psychology or sociology, or in any of the other natural sciences including physical geography and geology, or to graduate from the School of Education with a major or minor in any of these fields."[89]

Commenting on this action, Professor Langer, of the University of Wisconsin, writes as follows:

> *It seems important under the circumstances,* firstly, *that the facilities for the study of the mathematical subjects in the high school be guarded against any impairment, and that under no circumstances they be allowed to deteriorate from their present status.* Secondly, it seems important that students be brought at an early stage to a realization of the rôle played by mathematics both as a subject in itself, and as a buttress to innumerable others many of which are not obviously related to it.[90]

Mathematics as a mode of thinking. It is true that our unitary, one-year courses in algebra and geometry give a totally inadequate picture of the "instrumental" values of mathematics. This is not the fault of the subject nor of the teachers, but of the insufficient time allotment. Besides, even the undeniable, practical contributions of mathematics have failed to impress certain critics. These writers readily admit the "ancillary" functions of applied mathematics in modern life. But they follow Professor Inglis in assigning the "direct" values of a subject to the domain of vocational or practical training, and concerning the "indirect" values of mathematics they continue to have their doubts.[91] This skeptical attitude as to the "general" values of mathematics is still due, to a very large extent, to the unfortunate "mental discipline" controversy. It appears

[89] Langer, R. E., "The New Mathematics 'Requirement' at the University of Wisconsin," *The American Mathematical Monthly,* April, 1935, p. 211.

[90] *Ibid.,* p. 210.

[91] Inglis, Alexander, *Principles of Secondary Education,* Chap. XI, Houghton Mifflin Co., Boston, 1918.

that the latest phases of this age-old debate have not reached those who complacently and erroneously prefer to believe that "mental discipline is a myth." They have not heard of the disproving research work of Lashley and Orata.[92] They continue to maintain a wrong position, to the detriment of our young people.

In the light of the latest psychological findings, secondary mathematics may now reassert its *"cultural"* claims with even greater confidence. These claims have been publicized ever since the days of Pythagoras and Plato. In fact, the potential training values of mathematics are assuming ever greater dimensions. Whether we are thinking of arithmetic, or algebra, or geometry, or trigonometry, in every case we are in the presence of essential modes of thinking without which modern life could not endure. The number system is our great instrument of *precision.* It underlies all forms of *exact measurement.* The shorthand of algebra represents a type of *symbolic* thinking that has invaded many other sciences. Again, geometry gave to the world the first and most universal form of *postulational* thinking, as a model that will never be outmoded. Lastly, trigonometry has led to an ever-expanding view of our Larger Home, opening up by *indirect measurement* the vast expanse of infinite space.

H. G. Wells has reminded us that "the new mathematics is a sort of supplement to language, affording a means of thought about form and quantity, and a means of expression, more exact, compact, and ready, than ordinary language."

Mathematics is sometimes defined as the science of *relationships.* It may also be called the science of *invariance.* Hence its all-embracing character. For, as Professor Keyser has said so beautifully, *"to be is to be related."* The same writer has also given us perhaps the briefest and most compelling statement of the *cultural* function of mathematical thinking. To quote:

> Every one knows that one of the outstanding facts of our world is the great fact of Change. The world of events, whether great or small, mental or physical, is a flowing stream. Transformation, slow or swift, visible or invisible, is perpetual on every hand. But events are interdependent so that change in one thing or place or time produces changes in other things and places and times. With the processes of change every human being must deal constantly or perish. The proc-

[92] See Lashley, K. S., *Brain Mechanisms and Intelligence,* University of Chicago Press, 1929; see also Orata, P. T., "Transfer of Training and Educational Pseudo-Science," *The Mathematics Teacher,* May, 1935, pp. 265-289.

esses of change are not haphazard or chaotic, they are lawful. To deal with them successfully, which is a major concern of man, it is necessary to know their laws. *To discover the laws of change is the aim of science. In this enterprise of science the ideal prototype is mathematics,* for mathematics consists mainly in the study of functions, and the study of functions is the study of the ways in which changes in one or more things produce changes in others.[93]

PART FIVE

MASS EDUCATION, VOCATIONAL TRAINING, AND THE PROBLEM OF STANDARDS

Preliminary statement. In 1880 there were 3,941,365 persons 14 to 17 years of age in the United States, of whom 110,277, or 2.8 per cent, were enrolled in public secondary schools. In 1930 there were 9,341,221 persons 14 to 17 years of age, of whom 4,354,815, or 46.8 per cent, were enrolled in public secondary schools.[94] At present, more than 50 per cent of our adolescent population are attending our high schools; and the hope has been expressed by some of the highest educational authorities of the country that in the near future we may have 100 per cent enrollment of all our young people.

Now, no large business showing an increase of nearly 4,000 per cent within a few decades should be expected to escape serious growing pains. When the depression was added to this situation, it brought an even greater influx into our secondary schools, while at the same time available funds were sharply reduced. Unemployment closed the door to early vocational opportunities. Many thousands of pupils found themselves up against "academic" curricula for which they were not ready. And so the stage was set for our present educational perplexities.

Two views of secondary education. The general public knows only of the impressive attendance figures of our secondary schools. These statistical evidences of growth have been given prominence in the daily press. We have built up by far the largest system of secondary schools in the world. More than any other country, we have opened the door of opportunity to our young people. A vast building program, resulting in huge and expensive additions to the

[93] Keyser, C. J., *Mole Philosophy and Other Essays,* pp. 93 ff., E. P. Dutton and Co., New York, 1927.

[94] See *National Survey of Secondary Education,* Bulletin 1932, No. 17, "The Secondary-School Population," p. 4, Washington, 1933.

school plant, has been the source of alternate pride and despair to the taxpayer.

But there is another side to the picture. Only a moderate beginning has been made thus far in ascertaining the real educational and vocational possibilities of those admitted to the high school, and in providing them with appropriate opportunities for growth. When tested by conventional scholastic standards, a very large percentage of the present high school population must be rated as "failures." It is impossible to deny or to ignore that fact. To make secondary education available to all who wish to apply for it is one thing; to maintain a reasonable degree of efficiency under such circumstances is quite another thing. The conflict between the demand for mass education and the preservation of reasonable standards constitutes the dilemma which democratic education is now facing.

Revelations of the Measurement Movement

The effect of educational tests and measurements. When the measurement movement was launched on a large scale in the field of education, about two decades ago, the opinion was expressed very commonly that a new era had dawned on education. We are now able to appraise the significance of that movement more sanely. Grievous mistakes were made by the pioneers. Influenced by the spirit of Thorndike's famous dictum that "whatever exists at all, exists in some amount," a veritable deluge of testing devices began to flood the market.

The revelations of the subject-matter tests that were soon tried out on millions of children were certainly disquieting. Soon there was added the alleged proof that a child's I.Q. is constant. School procedures everywhere were profoundly affected by these findings. An educational determinism of unparalleled intensity threatened to sweep aside our brave optimism of former years.

The follies and scientific inaccuracies of many investigations in the field of educational measurements are now admitted by all who are interested in facts rather than fanciful theories. Ever since Professor Bagley's heroic battle against the erroneous assumptions and implications of the testing movement, the light of a clearer understanding has continued to spread.[95] We no longer expect to

[95] See Bagley, William C., *Determinism in Education,* Warwick and York, Baltimore, 1925; Rugg, H., "After Three Decades of Scientific Method in Education," *Teachers College Record,* November, 1934; see also Orata, P. T., "The Insignifi-

solve our major educational problems by the naïve expedient of "counting noses."

In the following pages we shall refer to certain educational tests not because we share the pessimistic interpretation commonly associated with them, but because we shall derive from them certain conclusions that are pertinent for the purposes of this discussion.

Educational records in English and mathematics. Since language and mathematics are the most highly sequential subjects of the curriculum, they are bound to reveal, more than any other subject-matter fields, the difficulties which both the elementary and the secondary school encounter in their effort to solve the problem of mass education. There is fairly conclusive evidence that English and mathematics go hand in hand in showing up the numerous maladjustments which have been caused especially by the lower ability levels in every modern school system.

With reference to reading, a very large amount of statistical material is now available. Thus, Assistant Superintendent John L. Tildsley of New York City reports as follows in the official monthly journal of the New York high schools:

> In June last we gave the Terman test to 27,573 pupils who entered the 9A grade of our high schools from public elementary and parochial schools. The range of I.Q. was from 55 to 174. Some 6,109 registered I.Q.'s of less than 90, 22.4%. We then gave the Haggerty Reading Test to the pupils who registered less than 90 in the Terman Test and learned that 87.5% scored less than the norm for the 8th grade. These results bolstered the judgment I had expressed before that we had between 40,000 and 50,000 pupils in our high schools who were not equipped to do high school work as hitherto conceived.[96]

An immense amount of work has likewise been done in ascertaining the degree of mastery achieved by pupils at the various grade levels in the field of mathematics. Thus, Professor Raleigh Schorling, of the University of Michigan, gave a test comprising 100 simple tasts in arithmetic to 3,545 pupils in grades 5 to 12 inclusive.[97] Some of the items in this test and the corresponding percentages of correct responses on these items in grades 7, 8, and 9 are reproduced on the following pages.

---

cance of the Statistically Significant," *Educational Administration and Supervision,* May, 1934, pp. 321-340.

[96] Tildsley, John L., "Planned Education," *High Points,* June, 1935, pp. 13-14.

[97] Schorling, Raleigh, "The Need for Being Definite with Respect to Achievement Standards," *The Mathematics Teacher,* May, 1931, pp. 311-316.

Questions Selected from Professor Schorling's List

Questions	Per Cent of Correct Responses Grade 7 638 Pupils	Grade 8 633 Pupils	Grade 9 335 Pupils
5. Add: $\frac{7}{8} + \frac{3}{16} =$ ____.	58.4	65.5	73.4
7. Add: $\frac{3}{5} + \frac{1}{2} + \frac{7}{10} =$ ____.	50.0	60.1	78.2
14. Add: $8\frac{3}{8}$ $5\frac{3}{4}$	45.7	52.2	57.3
19. Subtract: $9\frac{1}{4} - 2.2 =$ ____.	22.4	51.4	59.1
23. Multiply: 769 708	54.5	58.8	69.3
24. Multiply: $\frac{3}{4} \times 60 =$ ____.	68.2	81.3	86.6
28. Multiply: $20\frac{3}{5} \times 12 =$ ____.	30.0	42.2	65.4
29. Multiply: $1\frac{1}{2} \times 2\frac{1}{4} \times \frac{3}{4} =$ ____.	31.7	38.6	40.6
30. Multiply: 4.928 3.2	32.1	40.2	52.4
31. Indicate the decimal point in the answer: 20 × .20 = 400	69.8	76.8	85.6
35. Indicate the decimal point in the answer: Does 1.2 × 0.5 equal 6.0 or .60 or .060 or 60?	46.5	64.5	67.7
40. Divide: 85) 343400	36.2	48.6	63.8
43. Divide: $2\frac{1}{2} \div 4\frac{1}{2} =$ ____.	38.1	40.0	46.5
48. Indicate the decimal point in the answer: .123) 560.88 (quotient 456)	3.4	16.2	27.2
52. Does 4786 ÷ 10 equal 4.786 or 47.86 or 478.6 or 4786?	9.5	17.7	31.6

Questions	Per Cent of Correct Responses		
	Grade 7 638 Pupils	Grade 8 633 Pupils	Grade 9 335 Pupils
55. Write the fraction as a per cent: ¾ = ——.	51.2	82.2	76.3
64. Write the following as a decimal fraction: ½ = ——.	56.2	74.7	79.0
67. Write the following as a decimal fraction: 40% = ——.	42.6	53.2	51.6
74. Write as a common fraction: 10% = ——.	39.1	48.7	62.4
79. 25% of 80 = ——.	42.7	50.0	57.5
83. ——% of 48 = 16.	9.6	12.8	22.0
88. 7 = ——% of 14.	12.2	18.2	25.2
91. What is the average of 4, 6, 8 and 10?	18.6	39.5	46.5
97. Complete: the interest on $150 for 2 months at 6% is ————.	4.6	20.4	34.6

The following table presents the norms for grades 5 to 12, inclusive, on the same 100 tasks in arithmetic.

	Grade							
	5	6	7	8	9	10	11	12
No. of Pupils	622	636	638	633	335	236	230	215
75 percentile	34.2	28.8	50.6	54.2	59.8	62.3	72.2	81.9
Median score	17.6	22.0	39.8	43.8	48.6	51.7	59.0	67.0
25 percentile	12.3	17.6	30.4	35.0	39.9	42.6	46.0	54.3
Lowest score	1.0	3.0	6.0	9.0	5.0	12.0	16.0	22.0
Highest score	31.0	70.0	86.0	91.0	94.0	92.0	96.0	94.0

The same investigator tested 2,693 freshmen in the University of Michigan on the fundamentals of arithmetic and algebra.[98] Some of the questions and the corresponding percentages of correct responses are reproduced on page 90.

An algebra test prepared by Mr. Harold Lundholm of the University High School, State University of Iowa, was given in 1932 to 11,572 pupils in the Iowa high schools at the completion of a course in ninth-year algebra. It was a power test in the sense that approxi-

[98] *Ibid.*, pp. 317-319.

SELECTED TEST ITEMS FROM PROFESSOR SCHORLING'S TEST

No.	Selected Test Items	Per Cent of Correct Responses
	Part I (Arithmetic)	
4.	Multiply 86.5 by 0	67
5.	What is the average of 4, 11, 66, 0, and 9?	71
11.	Reduce 3/20 to a decimal fraction	68
15.	Write as a decimal fraction 399 millionths	45
16.	What per cent of 25 is 75?	28
20.	What per cent of 2 is ¾?	14
	Part II (Algebra)	
1.	If d is the cost per dozen, what is the cost of one?	81
5.	What is the square of $(x + y)$?	74
7.	Perform the multiplication: $-3x(2 + 5y)$	64
10.	Add: $\frac{1}{a} + \frac{1}{b} =$ ____.	27
12.	If $cy + d = 0$, and c is not equal to 0, what is the value of y?	12
13.	What are the factors of $a^2 - b^2$?	74
15.	Add: $\frac{1}{m} + n =$ ____.	19
16.	Add: $\frac{1}{x} + \frac{1}{y} + \frac{1}{z} =$ ____.	24
17.	For what value of x is the expression $\frac{x}{x+1}$ without meaning?	.023
18.	Simplify: $\sqrt{(64)(81)a^2b^{16}}$	14
	Part III (Presumably a "Power" Test)	
1.	Write as a power of 12 the number of cubic inches in a cubic foot	34
2.	The perimeter of a square is $4y$. What is the diagonal?	21
6.	What is the difference between the squares of two successive integers, the smaller of which is y?	15
7.	The area of a triangle is ab. What is the product of the base and altitude?	34
8.	Write a formula which gives the cost c in dollars for m miles of wire at d dollars a pound, if one foot of wire weighs p pounds	20
	(Write equation only)	
13.	Find two numbers whose sum is 100 and whose difference is 28	30

mately one minute was allowed per question. The test was of the "objective" type and the items selected were of a comparatively simple nature. The pupils were not ready for the prompt reactions expected of them. In Part I, comprising twenty questions on algebraic relations and representation, the percentages of correct responses fell below 50 per cent on eleven items. In Part II, containing twenty questions on the fundamental processes, fourteen questions scored below 50 per cent of correct responses. Part III contained work in problem analysis. Of the fifteen items included, fourteen fell below 50 per cent of correct responses.

In his *Psychology of Algebra*, Professor Thorndike reported on a test comprising twenty-eight simple algebraic tasks which he gave in ten schools that had a high educational rating. All the pupils tested had studied algebra for at least one year. The results showed an appalling weakness all along the line. In the language of Professor Thorndike: "It does not seem an exaggeration to say that, on the whole, these students of algebra had mastery of nothing whatsoever. There was literally nothing in the test that they could do with anything like 100 per cent efficiency."[99] Findings such as these led Professor Thorndike to express the opinion that pupils having an intelligent quotient below 110 are, in general, unable to understand the symbolism, generalizations, and proofs of the usual course in algebra. He also stated that if this conclusion be accepted, it would rule out more than 56 per cent of the present first-year algebra students. He recommended that pupils who are thus excluded might be given a course in mathematics within their capacity, or, in some cases, might study the customary first-year course in algebra during their second or third year.

We have given some of these figures merely to explain why many educators seem to feel that the traditional work in mathematics as at present organized is too difficult for the majority of elementary and secondary students and that for this reason alone mathematics should be made an elective subject. Again, they are of the opinion that if mathematics is to be retained at all in the general course, it should be so "simplified" that pupils of average ability can pursue it with a higher degree of measurable success. We shall examine these arguments at a later point in this discussion.

Adapting the High School to Individual Differences

Attempts at adjustment. In the light of facts such as those mentioned above, the high schools have tried, often very earnestly, to find a way out of the present impasse. They have attempted many adjustments or modifications of their traditional procedures.

In 1932 the Committee on Secondary School Problems of the Associated Academic Principals of New York State, as a result of

[99] Thorndike, E. L., *The Psychology of Algebra*, pp. 320 ff., The Macmillan Co., New York, 1924.

five years of study, adopted nine theses as an expression of the philosophy of education to be followed in that state.[100] "The first of these theses stated that secondary schools should provide education adapted to the needs of all pupils. This thesis recognizes that in the modern high school there are enrolled pupils of every degree of ability and interest, that the idea of selectivity is given up, that the single standard of achievement is abandoned, and that a high school education is the right of every normal boy and girl."[101] The association devoted its annual session, December 26-28, 1935, to a discussion of the question of "how this philosophy can be put into practice."

We touch here on one of the most vulnerable spots in our entire educational program. It has long been regarded as a settled issue that "every *normal* boy and girl is entitled to a high school education." But what about those who are *not* "normal," who have *not* mastered even the *rudiments* of an elementary education? The answer now commonly given is that for them, likewise, we must provide some form of secondary education which meets their "special needs." We can hardly do otherwise. To neglect these boys and girls would be suicidal. But from that point on the roads diverge very sharply. To continue to throw these backward and unprepared pupils into the ordinary high school courses is to invite the complete collapse of both secondary and higher education.

At present the high school is merely a "receiving station." As one principal has expressed it, *"Everybody goes to high school. The genius, the normal, the halt, the lame, and the blind come to us together and on equal terms."* Shall we not at last apply the principle of individual differences, so often invoked for sentimental purposes, *in a really scientific way?* What is the use of all our testing and measuring if we refuse to do anything about the facts which they constantly bring to our attention?

Administrative attempts at adaptation. On the surface, a plan involving mass education would seem to preclude the possibility of emphasizing individual differences in the classroom. When the number of pupils assigned to a class exceeds thirty, and when the teacher carries a load of five or six such classes a day, it is impossible for the teacher to keep in mind the individual interests and

[100] *Summary Report of the Committee on Secondary School Problems,* p. 16, Associated Academic Principals, 1932.

[101] *New York State Education,* December, 1935, p. 248.

needs of the pupils enrolled in her classes. When, in addition, the teacher must follow a prescribed curriculum, and when her pupils are expected to face uniform city-wide or state-wide tests, it is evident that the dice have been loaded heavily against the idea of adaptation to individual needs. In proof of this statement the table on page 94 is reproduced from Bulletin, 1932, No. 17, of the National Survey of Secondary Education.

An inspection of the table reveals that *not a single one of the many plans reported as being in use may be regarded as satisfactory in any real sense of that term.* It will be observed that only a moderate number of schools carried on "remedial" work or depended on "adjustment" classes, and that the schools reporting these plans as successful represent a comparatively small fraction of the total number of schools included in this study. Evidently, then, we must look in a different direction for a solution of the problem of mass education.

Subject-matter adaptations. The testing movement, as already suggested, has clearly proved the existence of a vast amount of retardation in the entire school population. From a standpoint of mastery, we simply have not arrived at any fair degree of success. With a resignation amounting to fatalism we have come to feel either that "nothing can be done about it" or that all requirements must be reduced to such a level of simplicity that even the slowest pupil can cope with them. The latter plan appears so plausible that the administrator tends to favor it because it promises to reduce the percentages of retardation and "failure." *Its tremendous danger, however, is in the direction of such a dilution of the curriculum for all pupils that very little of value remains for anybody.* Thus, speaking of algebra, one experienced administrator said to a group of teachers of mathematics: "Unless you can bring this subject down to the level of morons, you need not expect to teach it very much longer." In other words, the "moron" is to determine the educational program of the school.

An examination of current textbooks shows conclusively the following unmistakable results of this demand for "easyism": (1) the vocabulary has been greatly simplified, in accordance with Professor Thorndike's efforts in that direction; (2) many "technical" details have been removed; (3) the entire presentation has been made much more elementary.

In mathematics the simplification movement has already achieved

Frequencies with Which Various Provisions for Individual Differences Were Reported in Use, or in Use with Unusual Success, by 8,594 Secondary Schools*

Provision	Use		Use with Estimated Unusual Success		Column 4 Divided by Column 2
	Number	Per Cent	Number	Per Cent	
1	2	3	4	5	6
1. Variation in number of subjects a pupil may carry	6,428	75	795	9	0.12
2. Special coaching of slow pupils	5,099	59	781	9	.15
3. Problem method	4,216	49	444	5	.10
4. Differentiated assignments	4,047	47	788	9	.20
5. Advisory program for pupil guidance	3,604	42	540	6	.15
6. Out-of-school projects or studies	3,451	40	439	5	.13
7. Homogeneous or ability grouping	2,740	32	721	8	.26
8. Special classes for pupils who have failed	2,612	30	350	4	.13
9. Laboratory plan of instruction	2,611	30	323	4	.12
10. Long-unit assignments	2,312	27	349	4	.15
11. Project curriculum	2,293	27	365	4	.16
12. Contract plan	2,293	27	465	5	.20
13. Individualized instruction	2,145	25	309	4	.14
14. Vocational guidance through exploratory courses	1,911	22	186	2	.10
15. Educational guidance through exploratory courses	1,900	22	193	2	.10
16. Scientific study of problem cases	1,343	16	146	2	.11
17. Psychological studies	1,077	12	70	1	.06
18. Opportunity rooms for slow pupils	946	11	172	2	.18
19. Morrison plan	737	9	175	2	.24
20. Special coaching to enable capable pupils to "skip" a grade or half grade	726	8	114	1	.16
21. Promotions more frequently than each semester	686	8	103	1	.15
22. Remedial classes or rooms	593	7	90	1	.15
23. Adjustment classes or rooms	544	6	55	1	.10
24. Modified Dalton plan	486	6	52	1	.11
25. Opportunity rooms for gifted pupils	322	4	69	1	.21
26. Restoration classes	191	2	24	0	.13
27. Dalton plan	162	2	15	0	.09
28. Winnetka technique	119	1	14	0	.12
29. Other	101	1	..	..	..

* From National Survey of Secondary Education, Monograph No. 13, U. S. Office of Education, Bulletin 1932, No. 17, Table 4, p. 9.

considerable dimensions. This is true of arithmetic especially, but also of elementary algebra and plane geometry. The latest development is a group of one-year orientation texts that have given up all attempts at coherence and thoroughness. In these books, a series of "snapshots" has replaced the sequential presentation of more substantial treatises.

Some cities have also begun to issue their courses of study in various editions, according as they are intended for good, average, or slow pupils. Grade-placement charts have been modified in a similar way. A three-track system is officially sanctioned in many schools, and a different time-schedule is in force for each of the three different levels of ability officially recognized.

Educational guidance. For some years, also, the schools have been doing a gratifying amount of "follow-up" work, scrutinizing with greater care the progress of the individual pupil. But for the depression, which curtailed the clerical and executive staff of many an institution, much more progress in that direction would now be in evidence. Again, with increasing frequency, the educational history of each pupil is being recorded, thus enabling the official advisers of the pupil to deal with him more wisely. At best, these counsellors have a very difficult task to perform. Often they follow the line of least resistance by sending the pupil from the "academic" classroom to the shop or to the vocational school. Such educational "guidance" may rescue the pupil from the Scylla of "failure," but quite commonly substitutes for it the Charybdis of a closed door at the end of the line. Other attempts at adaptation to individual differences, such as the introduction of differentiated courses and of special adjustment programs, will be discussed at a later point.

The Issue of Vocationalism

Statement of the issue. The problem of the unadjusted or retarded high school pupil has led many to advocate a solution which at first sight looks very promising. It is to the effect that the majority of these young people are not "book minded," that, instead, they are "hand minded." Hence they should be sent to vocational schools offering courses in the practical arts, in trade instruction, in pre-engineering work, and the like. It is asserted that the real trouble with the "academic" type of high school is that it lacks the "life-career motive" and therefore does not appeal to our most active and vigorous boys and girls. Once more we meet here the old argument of narrow and early vocationalism against general training and liberal education.

Prevailing misconceptions and fallacies. We have already pointed out the importance of adequate hand training and of art appreciation for every boy and girl. But we must carefully distinguish such liberalizing contacts from a strictly vocational emphasis. What, then, are some of the pros and cons of vocationalism at the secondary level?

1. If it were possible to diagnose correctly the latent practical or professional abilities of each fourteen-year-old pupil, and then to plan and to provide for each an appropriate course of study, many of our troubles would probably vanish. We are far from being able to carry out such a scheme. Most children are not sufficiently mature at the age of fourteen or fifteen to feel reasonably sure about their permanent life interests. *A premature vocational choice would lead to misery and social maladjustment.*

2. It is not true, as one prominent administrator asserts, that "eighty per cent of the high school pupils are hand hungry." The majority are not. Even a superficial acquaintance with shop schools serves to bring out the fact that they complain even more bitterly about their unadjusted and aimless pupils than do the "academic" high schools. The lower grade of trade schools have become the "dumping ground" of "problem cases" and often resemble reformatories rather than "joyous laboratories of direct experience." In other words, "handwork" alone is not the glorious miracle formula though theorists cling to it ardently.

3. At best, trade schools of this type furnish an opportunity for apprenticeship training. They cannot turn out master craftsmen. It is true that the graduates of such schools are still in demand in large industrial cities. In prosperous times they can usually find immediate employment as junior craftsmen. But if they do not continue their education, they often receive a rather unsatisfactory wage and also face the certainty of periodic or permanent unemployment before they reach middle age.

4. Our industrial system, as everybody knows, is in such a state of turmoil at present that it takes more than courage to recommend a narrow type of industrial training to any youngster. Today, technological unemployment predominates. Its problems demand a much broader educational equipment. Hence *those who know a*

good deal about modern industry are advising against early industrial education.

5. It is highly significant that industrial and agricultural workers as a group are strongly opposed to a merely "practical" education for their own children. They know something of the limitations resulting from an exclusive emphasis on handwork. Again, the young people who crowd our evening high schools, after a day in the factory or in the office, know even more of the boredom of this machine age and have experienced the emptiness of mere manipulation.

6. So mechanical has become our way of living, so enslaved are we to the machine and to the routine of business, that only a broadly trained mind can hope to find in the rich cultural possessions of the race an antidote to the weariness resulting from the daily grind, and a satisfying refuge for leisure hours. It seems almost criminal to rob young people of the few precious years which they have to get ready for the wear and tear of life, by a premature injection of the mechanized performances which only too often are destined to become their steady diet.

No royal road to learning. It is also a fatal error to suppose that early vocationalism somehow corrects or makes less serious the scholastic or attitudinal shortcomings of the pupil merely by supplying a more potent or more direct kind of *motivation.* A stenographer or typist who cannot spell is discharged or relegated to a secondary place. A machinist who is unable to measure accurately and to make shop computations cannot expect to hold a steady job. Everywhere the demand is for greater efficiency, for good judgment, for almost instantaneous reactions. A dawdling, shiftless boy is simply "out of luck" in a modern industrial plant or in almost any practical career. At best, he may look forward to the occasional and uncertain employment accorded to the unskilled laborer, since in effect he is a mere "hand."

How, then, is the vocational school attempting to supply that mastery of the fundamentals which even the humblest industrial or commercial worker simply *must* possess? An inspection of many shop texts and "practical" courses of study shows that only too often the situation is met by ignoring it. Thus, in mathematics, no attempt is made to develop *insight.* The fundamental processes are "reviewed" by a lot of mechanical, imitative work which is based not on principles but on unexplained rules. There is no evidence of acquaintance with modern educational objectives, or with the newer psychology of learning. Obsolete processes and outmoded types of exercises are continued in abundance. Not infrequently, the organization of the topics and the presentation of new materials are almost archaic.

Teachers of shop classes have found that it is next to impossible to carry on group instruction when no two pupils in a class have a common educational background. The disabilities of such pupils often date back five or more school years. As a result, shop teachers commonly depend on individual instruction, with the aid of lesson sheets, special shop manuals, or work books. With thirty-five pupils in a class, each individual pupil may receive as much as five minutes a day of individual instruction. At that rate, backward pupils will sometimes take many days to complete a single lesson. Such is the price we pay for the mistaken notion that "vocational" training, instead of foundational teaching, is a quick remedy or a cure for educational retardation.

The old and the new conception of vocational education. Educational treatises written thirty or forty years ago assign to the secondary school the task of direct life preparation. Professor Hanus insisted that it is the function of such a school "to stimulate every individual to aim at intelligent self-support or some worthy form of life-work, whether he inherits an income or not; and to give him general preparation for such activity."[102] In his opinion the secondary school in the past prepared its pupils too exclusively for the professions, and:

> What we do not yet fully recognize is the function of the secondary school as regards the *vocational* aims of those who subsequently devote themselves to *industrial* and *commercial* pursuits. This function deserves recognition on the broadest grounds, both for the sake of the vocational interests themselves, and for the sake of all the possible interests which the individual or society has.
>
> This means that in addition to the purely intellectual courses of the school *we should maintain in every secondary school*, whether public or private, *courses in manual training, and commercial courses, which, together with their general educational aims, minister directly to vocational and social aims.*[103]

It is noteworthy, however, that even in the days of Professor

[102] Hanus, Paul H., *Educational Aims and Educational Values,* p. 82. Quoted by permission of The Macmillan Company, publishers. New York, 1902.
[103] *Ibid.,* pp. 85-86.

Hanus there was no intention to identify "manual training" with the teaching of a trade. One of the writers of the same period expressly says that "the working boy of today needs not so much any one trade as that combination of qualities which will enable him to turn with facility from one occupation to another as each in turn is supplanted in the course of the industrial evolution."[104] And commercial education was conceived so broadly that its ultimate aim, for the young man at least, was described as that of making him "a public-spirited, intelligent, well-educated and successful man of affairs."[105]

The "cosmopolitan" and the "manual training" high school carried on for years in the spirit of that tradition. For a while everything looked very harmonious. But the ever-increasing momentum of the machine age, the effects of technological unemployment and of economic insecurity, and the resulting influx of pupils into the high school, caused a gradual, yet inevitable change of orientation with reference to vocational education. We are now in the midst of a transformation that cannot be evaded or postponed. The Smith-Hughes Act of 1917 proved as ineffective as did the numerous "exploratory" shop courses that we still find in so many junior and senior high schools.

The new outlook is governed by the thesis that, in the future, *vocational education will be essentially a post-secondary problem*, and, accordingly, *the secondary school should limit itself primarily to general types of education.*

Who are the sponsors of this new view of vocationalism? Those who have been in close touch with its development and who have made thorough studies of our present industrial system.[106] Among the leaders in this investigation is Professor David Snedden of Teachers College, Columbia University. Whatever one may think of his drastic views concerning the *cultural* work of the high school, the many years of careful attention which he has given to the social and vocational problems of American schools compel us to heed his

[104] From "The Working Boy," by Florence Kelley, *American Journal of Sociology*, II., p. 158. Quoted approvingly by Professor Hanus, *op. cit.*, [102], p. 87.

[105] *Ibid.*, p. 90.

[106] To Professor Paul R. Radosavljevich of New York University the writer is indebted for an illuminating account of corresponding recent developments in Russia. Of particular interest in this connection are Professor Paul P. Blonsky's two volumes entitled *Trudovaya Shkola* (School of Work), Moscow, 1929. Professor Blonsky is the head of the largest school in Moscow (2,000 pupils, from the kindergarten through the second year of the secondary school), as well as a professor of philos-

authoritative advice in the field of vocational training. He has outlined his position in a series of recent publications.[106a] *He rejects in toto, as futile and inefficient, our present provision for vocational training at the secondary level,* including the usual "shop work" in junior high schools, as well as technical, manual arts, and industrial high schools, commercial departments, home economics departments, agricultural high schools, and Smith-Hughes vocational courses. He characterizes as "fantastic" the notion that junior high schools can profitably offer genuine preparation for any vocation. We are told that because of prevailing economic conditions, any one hundred boys now entering any typical American high school at fourteen or fifteen years of age will be found, fifteen years later, distributed among seventy to eighty unlike vocations. Professor Snedden predicts, somewhat prematurely, that *by 1940 all commercial departments will have disappeared from American high schools.*

Professor Snedden insists that in the future all pupils must attend full-time schools of *general education* until the end of their eighteenth year. Training for a junior vocation may begin at *eighteen*; for an operative vocation *not under twenty*; and for a mastership vocation *not under twenty-five.* He advocates "a New Deal system of full-competency vocational education *at the post-secondary level,*" consisting of one or a few schools specialized for each vocation in each populous state, and administered on a state-wide basis:

> Under the Expected New Deal a fully developed, fairly democratic system of vocational schools designed to serve all of the million young men and million young women who in these United States must each

---

ophy in the *Academy of Social Education,* called "Krupskaya." The following passage from Professor Blonsky's work (p. 164) is of great significance:

"We do not want to produce an artisan; the aim of the School of Work is not the teaching of a trade, but the development of the worker and brother and ally to all toiling humanity; a keen observer of the life of mankind, a ripe individual, yet retaining the altruism and poetic feelings of childhood, an individual living in close touch with present-day technique and science, nature and art—such is the product of the School of Work, bringing together active men and humanized nature; such is the little industrialist entering contemporary life by way of the historical evolution of man and in accordance with the evolution of childhood; such is he, the able and informed builder of the future of a better human life, mightily steeped in light and happiness."

[106a] See especially Snedden, D., *American High Schools and Vocational Schools in 1960,* Bureau of Publications, Teachers College, Columbia University, New York, 1931; see also "An Expected New Deal for Vocational Education," *Teachers College Record,* October, 1934, pp. 53-61.

year, somewhere, between their eighteenth and twenty-fifth years, enter upon full-time, sustaining vocational work, or, later, advance to higher vocations or stages of vocations, it is predicted: (*a*) will require perhaps from one thousand to three thousand distinctive kinds of vocational schools for the entire country; (*b*) will require altogether not to exceed fifteen to twenty thousand distinctive schools for the entire country; (*c*) will entail an annual cost not to exceed two hundred to four hundred million dollars per year; (*d*) will prove exceedingly simple of administration, once basic guiding principles are accepted; and (*e*) may be expected very soon, even in strictly pecuniary terms, to expand the productive and taxpaying powers of our population far beyond the annual cost of the system of vocational schools itself.[107]

The Problem of Standards

A battle of conflicting opinions. The writer has had the benefit of many group conferences with a large number of elementary and secondary teachers representing a considerable number of communities. He has never met one subject-matter teacher in the higher grades who failed to insist that without a reasonable foundation and without careful sequential training a pupil is almost certain to become a "problem case." Thus, a pupil cannot learn to multiply until he can add. He cannot solve percentage problems unless he can handle decimals. He is helpless in algebra without some proficiency in arithmetic. Teachers everywhere regard it as axiomatic that progress is impossible without a moderate amount of effort and self-activity on the part of the pupil.

When such abilities and attitudes are lacking, they cannot be created overnight by a sort of magic. The correction of deep-seated disabilities is never a matter of a few "remedial" lessons; very often these shortcomings call for heroic measures extending over weeks and months.

Now, the educational theorist seems to think that all these teachers are sadly mistaken. They have no right, we are told, to call any pupil a "failure." It is too discouraging to children to be branded in this way. Children should always have a buoyant sense of "success." Even though they misspell many common words, cannot read elementary texts, and cannot solve the most ordinary problems in arithmetic, these deficiencies should be no hindrance to a glorious, unimpeded development onward and upward.

[107] *Teachers College Record,* October, 1934, p. 58. See also *High Points,* June, 1935, p. 18.

At any cost, children *must* be promoted. It is the teacher's business to find out what can be done about it. Teachers have come to feel that etiquette forbids the mentioning of such an unpleasant fact as retardation. A conspiracy of silence is expected to eliminate all problems of maladjustment. Unfortunately, mathematics refuses to be built on anything but an honest foundation. It is incorruptible. Two and two will never make five. Nor can mathematical skills be acquired without careful, continuous effort over a period of years. Because of this refractory nature of the mathematical curriculum, it is now to be punished by being destroyed. The grand discovery has been made that we do not need anything but the barest rudiments of arithmetic in everyday life.[108] Everything else, including fractions and decimals, may safely be assigned to "technical" course at the higher levels. Such details, it is claimed, can readily be "picked up incidentally" whenever they are needed. This assertion that mathematics and the other fundamentals are superfluous or outmoded for the average pupil has been received with such warm approval by a group of "progressive" educators that one wonders how far this aberration will extend.

From the standpoint of the teacher, the disturbing and amazing thing about this development is that *no suggestion is ever offered as to how it is to be made possible to do any kind of educational work with a large group of pupils when no two of them have a common foundation.*

Let us now hasten to assert that *the real question which refuses to vanish from view is whether society has a right to expect from every child a reasonable mastery of the fundamentals in the trunk line subjects. If that question is answered affirmatively, as it should be, it will then be necessary to investigate the reasons for the educational disabilities of so many children and to find out what can be done to correct them.*

The rejection of standards. How far some educators are ready to go in their policy of indiscriminate adaptation to individual differences, may be inferred from the following statement, made by Professor Guy M. Wilson:

> Provision for individual differences, which is really functioning, means that *there are no failures,* no money spent on repeaters. A child is entitled to six years in the grades, three years in the junior

[108] Wilson, Guy M., *What Arithmetic Shall We Teach?* Houghton Mifflin Co., Boston, 1926.

high school, three years in the senior high school. Admission requirements for grade 1 is age 6, to the junior high school age 12, to the senior high school age 15. These are the general admission requirements: by agreement with pupil and parents slight modifications may be made in either direction. *The teacher's problem becomes one of adaptation. In all of the grades, the work is adapted to the interests and ability of the child.* Under this scheme college entrance as such is ignored by the public schools. The service viewpoint is developed.

The best single adaptation is that of granting to the teacher full freedom to disregard or *adapt administrative machinery to the needs of her particular children.* The teacher's major instruction should be: "Keep the child in mind; meet his real needs with sizable tasks, self-chosen, self-planned, and adapted to ability."[109]

At a recent educational meeting in the New York area, Dean Milton E. Loomis of New York University is quoted in the metropolitan press as having made a vigorous attack on "regimentation." He is reported to have said, in part:

If democracy is to survive, American practice must conform to American theory. We must free our institutions from the caste spirit, and *individualized education must lead the way.* . . . Adequate in their own day, the three R's are no longer sufficient or of paramount importance in modern educational technic. . . . The social studies and the fine arts are now the proper foundation of the curriculum.

This sudden and meteoric solicitude on behalf of the social studies and the fine arts fits in beautifully with the prevailing dislike of "high-mortality" subjects and the rejection of standards. One more quotation of this type must suffice. In the *Chicago Schools Journal* of September, 1934-February, 1935, there appears this passage:

Fortunately, the policy of reorganization is simple. *The basis for all educational reorganization should be the needs of society and of the individual for the present age, the present year, the present hour, the present minute.* Nearly all other ages have recognized this truth but we have preferred to follow a pattern cut for and by men of a day long since passed. The earliest civilizations insisted upon this principle of educating for the present. . . .

For the ninety-five per cent, however, who will not go to college, who lack the ability to succeed in college, and who should not go there in any case, different types of education should be offered. . . . If you will study without prejudice the curriculums of elementary and high schools throughout the country you will discover an amazing

[109] Wilson, Guy M., "The Child as the Center," *Second Yearbook of the Department of Classroom Teachers,* p. 13, Washington, 1927.

number of subjects that have little or no practical or cultural value for ninety per cent of the students.

The substitution of a new program for the old in the secondary schools will not be easy, but it must be made. Beyond art, music, some of the social subjects, and a limited amount of science the foundation is weak. . . . *To circumvent future depressions the study of economics at various levels will be demanded.* It can be made interesting even to young children, but at present it is ineffective. The stigma which the schools have so often placed upon labor will be removed and *vocational education, or education for versatility will become respectable.* The primary function of free public education in a democracy, the development of citizenship, will be recognized as the task of the public schools, from kindergarten to college. Guidance, direct and indirect, will lighten the pathway of adolescence and youth.[110]

Appraisals of the present trend. Fortunately, not all of our educational thinkers have failed to recognize the ominous implications of these tendencies. Professor Boyd H. Bode is of the opinion that the net result of current educational policies is that "our conception of general education has become a collection of odds and ends for which it is impossible to have any profound respect."[111] He says that "we teach a little of everything," expecting the pupils to achieve a synthesis in some unexplained way. Instead, "the various elements in their education tend to neutralize one another, and so the final result is apathy or intellectual and emotional paralysis."

The most searching critique of the principle of "adaptation," however, is that of Professor I. L. Kandel. In his Inglis lecture, entitled "The Dilemma of Democracy," he exposes with relentless logic both the fallacies and the fateful consequences of the educational theories we have been considering. This essay should be read by every teacher who is interested in the present situation. Professor Kandel points out the error of supposing that all subjects are of equal educational value. He traces this error to the convenient but totally false belief that the doctrine of generalized training, incorrectly called formal discipline, has been disproved. Recent investigations, as was stated previously, point in exactly the opposite direction. In brief, Professor Kandel finds the root of the trouble to be the constant confusion of freedom of opportunity with

[110] Bogan, William J., "Reorganization in the Public School System," *Chicago Schools Journal,* September, 1934-February, 1935.

[111] Bode, Boyd H., "The Confusion in Present-Day Education," in *The Educational Frontier,* Chap. I, Kilpatrick, William H. (Editor), [20].

equality. We submit the following significant passages from his monograph:

> *To insist that courses should be so differentiated as to be adapted to the needs and abilities of the pupils is one thing; to adapt standards to the abilities of the pupils is another and means the complete abandonment of any standards at all.* The significance of this principle is that high-school graduation is again reduced to the time element and scrambling through the required number of units at a pupil's own level. . . .
>
> *In the desire to do justice to all it is highly probable that the new dispensation will result in doing justice to none. In the confusion of equality with identity there is a very serious danger that the real problem is evaded of devoting that care, which is essential, to the discovery of ways and means of treatment most appropriate to individual differences of capacity.* In the widespread concern that pupils at the lower end of the scale of intelligence now admitted to the high school shall profit from their stay in school and that the word "failure" should be eradicated from school practice, questionable justice may be done to this group by the warping of standards to meet their capacities; but at the other intellectual level there is no doubt that injustice has been and is being done not only to the brighter pupils but to society as a whole. . . .
>
> The extension of the period of education for all adolescents is inevitable; it is not beyond the scope of wise educational statesmanship to devise the most appropriate methods of promoting equality of opportunity, equipping all with a common language of intercourse, giving to each according to his ability, and leaving an open road to talent.[112]

Some common misconceptions. In the current educational literature one constantly finds statements which are sources of misunderstanding. While it is certainly true that not all children can reach exactly the same goals, in a given period of time, is it not absurd to argue that because of this fact there should be no standards whatsoever? Again, those who advocate a multiplicity of standards never indicate how a school system can operate on such a basis. Finally, it is forgotten that "in man, the things which are not measurable are more important than those which are measurable."[113] In other words, many important types of training are not of the verbal type and cannot be tested by written examinations. There are physical, mental, and spiritual adjustments which demand constant attention in the school and which can never be recorded in terms of a standard scale.

[112] Kandel, I. L., *The Dilemma of Democracy*, Inglis Lecture, 1934, pp. 50-79, Harvard University Press, Cambridge, 1934.

[113] Carrel, Alexis, *op. cit.*, [27], p. 278.

Obviously, however, there are other aspects of the pupil's life in school which *do* require the application of definite and even uniform standards. The moral ideals of life are not determined by personal likes or dislikes. Honesty, truth, justice, integrity, loyalty, and the like cannot be juggled to suit the convenience of the crowd. Nor can intellectual and aesthetic standards be "adjusted." There are many situations which require the unflinching application of such standards.

We can sum up these apparent or real contradictions by saying that a pupil should certainly not be rated a "failure" so long as he is really trying, to the best of his ability, to improve in every possible way, within the limits of his capacity. On the other hand, *standards of excellence are never to be given up.*

In the light of these comments it would seem superfluous to point out in detail the mixture of truth and error in the solution recently proposed by Dean M. McConn of Lehigh University:

> It is easy to see now that the gospel of standardization was based in part on a tacit, uncriticized, and unwarranted assumption: the assumption, namely, that all men and particularly all children are equal and alike, or nearly equal and nearly alike, not only in their right to Life, Liberty, and the Pursuit of Happiness, as the democratic doctrine declares, but also in kind and degree of intelligence and capacity. In short, we quite overlooked the little matter of Individual Differences—of which, in fact, little or nothing was heard thirty or forty years ago.
>
> We entirely missed the fact that the great majority of the children in schools, and even a substantial minority of the undergraduates in colleges, were not at all like us, but were endowed with quite other kinds of capacity and often with lesser degrees of capacity of any kind.
>
> For those others—the majority!—our generalized uniform Standards were all wrong, in that they gave exclusive sanction and exclusive prestige to tasks which were unsuited to their kinds of capacity or impossible for their degrees of capacity or both. As a consequence, the Standards have caused, and are now causing, untold damage and untellable misery to vast numbers of children in the elementary schools and high schools and even in colleges, thwarting and warping and beating down young lives.
>
> We vaguely imagined that in providing and enforcing substantially the same kind of instruction for all children we were serving the democratic principle already cited: Life, Liberty, and the Pursuit of Happiness for everybody. But in fact our uniform and exclusively intellectual Standards have deprived a majority of our pupils of the last two of those rights.
>
> Plainly, then, *what we need is more standards; many highly differ-*

entiated and carefully graded standards, adapted to as many kinds of capacity and to as many levels of attainment as we can identify in the children actually in our schools. Each of the new differentiated standards would naturally—like our present Standards—carry its appropriate prescription or indication of subject matter or kind and method of instruction and its own norms of excellence. All should be given equal sanction, and to each should be accorded its appropriate prestige. Thus, and thus only, shall we succeed in bringing to all children those benefits—first-rate facilities and feasible goals and successful and happy attainment—which our old Uniform Standards sought to bring to all but have actually brought only to one limited group, namely, those who are in some degree bookishly superior.[114]

In defense of standards. In concluding this part of our discussion, we submit a few significant comments relating to the necessity of maintaining standards and to the danger of ignoring them. Professor Bagley has repeatedly summarized the consequences of planless and effortless education. From a paper entitled "How Shall We View Elementary Education?" we quote the following passage:

We turn now to subject-matter. . . . There is no term—not even discipline—that is in greater disrepute in educational theory and practice today. The severest epithet that one can hurl at a teacher or an administrator is to say that he is "subject-matter conscious." I have not discovered what the antithesis, or corresponding virtue, really is. If "subject-matter consciousness" is the vice, is unconsciousness of subject-matter the virtue—or perhaps innocence of subject-matter—since innocence and virtue are closely associated? I am left myself in the predicament of the student of the new physics; I cannot comprehend education without subject-matter;—but perhaps I could if I had command of the proper symbolism. . . . *The slightest reflection is sufficient to prove that an enlightened system of education must in large part reflect authority. It should not reflect an authoritarian control by vested interests, but it should reflect the authority of tested human experience. . . . It is the background of experience that determines the meanings which we read into the impressions that impinge upon consciousness. . . .*

The systematic and sequential mastery of past experience as organized in the various fields of human inquiry I regard as the most dependable source of helpful backgrounds. It is easy to deride such mastery as the accumulation of mere information; it is easy to discredit learning processes that demand effort and sustained attention in the face of desires and distractions. One may be sure that any proposal which sanctions and rationalizes the lines of least resistance

[114] McConn, M., "Examinations Old and New: Their Uses and Abuses," *The Educational Record,* October, 1935, pp. 7 ff. Reprints may be obtained from The American Council on Education, Washington, D. C.

will receive the applause both of the gallery and of a goodly section of the pit.[115]

Writing on "the fallacy of the passing grade" and our customary "get-by attitude," Professor Henry C. Morrison of the University of Chicago states the situation as follows:

Study of problem cases, in the laboratory schools especially but elsewhere as well, seems to show in many pupils a characteristic well-defined volitional perversion which we have come to call the "get-by attitude." The pupil thus afflicted—and the victims are many—comes to see any task which he has to do, not as a thing to be accomplished in a finished manner as a matter of course, but rather as an undertaking upon which he will economize effort to the degree which experience has taught him will be accepted. If we raise the standard by requiring a higher passing grade for certain purposes, we simply require greater exertion without changing the attitude. As the pupil goes on into high school and college, he often becomes very skilful in his ability to just scrape through. What college teacher has not met these people? They occasionally become solicitous about grades and upon being assured that their work is acceptable, receiving B perhaps when the passing grade is C, they at once relax.

Now, the teacher, and perhaps the parent, passes this all off with a good-natured smile and the comfortable verdict "Just a boy—he will come out all right." Yes, it is just a boy—a nine-or-ten-year-old boy. *The truth of the matter is that the attitude when found in high school or college is a serious perversion and, unless corrected, results in permanent volitional retardation. The attitude carried over into adult life means irresponsibility, low standards, and, whenever the social controls become relaxed, lawlessness in a variety of social relations. Such an adult is incapable of becoming a citizen, in the social sense of that term,* albeit he may legally be capable of voting and holding office.[116]

Assistant Superintendent John L. Tildsley of New York City is substantially of the same opinion. He says:

In view of the experience to which our boys and girls have been subjected in our schools of the generation past, the ever lowered standards, the ever lessening demands made upon them, the failure to hold them to even the lowered standards, with this effect heightened by the example set everywhere in our parks and on the streets by men on relief or in the regular City service, *I am convinced that the most important change we need to produce in these boys and girls in our high schools is to build up in them as a lasting possession, the urge*

[115] Bagley, William C., "How Shall We View Elementary Education?" *The Mathematics Teacher,* March, 1935, pp. 170-180.

[116] Morrison, Henry C., *The Practice of Teaching in the Secondary School,* pp. 45 ff., University of Chicago Press, Chicago, 1926.

to do one's best. I would be willing, if it were necessary, to exchange for possession of this quality, knowledge of many subjects and many skills. To the boy possessed of this asset, knowledges and skills inevitably are added. In fact, in the very act of developing this quality, knowledges and skills accrue as by-products. *This quality has been through the ages the dominant quality of the great artist, of the gild craftsman, even to our own day.* It was rather common in the schools of my boyhood. It was common among the New England housewives. Each sought to make the last pie the best ever. *Today in the high schools of New York it is most uncommon.* It reigns chiefly in athletics, especially in sports requiring team play. Not to do one's best when on a team is taboo. . . . *Something more than interest is required.* Many educators to the contrary, I am firmly convinced that in a school all of whose teachers are permeated by this urge, it can be in large measure universalized among the students to the extent that in whatever a student undertakes that he thinks worth the effort, *this urge which has become a vital element in his make-up will drive him on towards the, for him, perfect performance.*[117]

We have not yet spoken of the taxpayer and the parent who are affected at least indirectly by the battle that is going on in the field of education. While parents, very naturally, prefer a "passing grade" for their children, they are by no means oblivious to the danger of purely complimentary marks. It is safer to tell them the truth. In the long run, they are bound to hear it anyway, often when it is too late. For some time, the dangers of "painless education" have received attention in the daily press and in the periodical literature. We quote merely the following passage from a recent paper by Professor E. A. Cross in *The Atlantic Monthly*:

Among the patrons of the public schools there has grown up a conviction that *something is vitally wrong with our system of education.* This feeling at times breaks into verbal expression as a vague dissatisfaction, and again as a more localized faultfinding with some particular phase of education, such as spelling, arithmetic, or handwriting. *Seldom does it go deep enough to touch the vital core of the malady.*

The American system of education ought to be the best in the world. It is universal. Every child not only *may* attend school, but is *required* to do so. The state does not discontinue free education at the end of the elementary period, but permits attendance in tax-supported schools, colleges, and universities up to any age. In buildings and material equipment our schools are not equaled by any in the world. We have the most elaborate and expensive system of public training schools for teachers. We pay our teachers, not enough, but certainly more than teachers are paid in other countries.

[117] *High Points,* June, 1935, pp. 16 ff.

But, with physical conditions as nearly ideal as we have them, *the parents and the taxpaying public are not satisfied with the results we get.* This dissatisfaction is not limited to critics who stand outside the organization and look on. *It is shared by many of the most thoughtful teachers working within the system.* They often see the faults more clearly than the parents, the taxpayers, and the newspaper critics can see them. *And thousands of teachers are willing and anxious to do something about these shortcomings. . . .*

The products by which we must judge the schools are the children who come out of them at the age of fourteen, or eighteen, or twenty-two, labeled by the system as educated young people. *One of the fundamental considerations offered to justify universal free education in a republic is that it creates thinking, law-abiding, and morally clear-seeing citizens.* A self-governing state cannot continue without such a body of citizens. *Do our schools succeed in turning out men and women of this kind? By the thousands, yes. But by the millions, no.*[118]

If these outspoken and honest appraisals be deemed inconclusive, attention may be called again to the most colossal educational experiment of all time. So great has been the revolt[119] against planless curricula and "progressive" experimentation that there is to be a return to extreme and deplorable regimentation. Thus, hereafter, all boys and girls are to wear uniforms identical throughout the Soviet Union for each type of pupil. Examinations abolished in 1917 have been revived. The authority of the teacher is to be restored. Experimentation is to be strictly limited. *Will this debacle at last cause our own visionary theorists to see the handwriting on the wall?*

The Supreme Necessity: An Integrated Program of Foundational Instruction

Summary of the underlying problems. What, then, it is now necessary to ask, can we do about our staggering problem of mass education? Before we proceed to outline plans that look more promising than those discussed above, let us state once more, in briefest form, the facts which must be faced and the conflicting ways in which educators have reacted to them.

1. *The evidence is overwhelming that millions of children are retarded in their school work.* Every scholastic test reveals far-reaching disabilities in the fundamental studies. Every public

[118] Cross, E. A., "Painless Education," *The Atlantic Monthly*, December, 1935, p. 740.

[119] See reference [49].

school teacher, especially in the upper grades and in the secondary schools of large cities, is aware of an alarming amount of maladjustment. The educational edifice shows signs of such strain that indifference to the situation seems perilous and out of the question.

2. The prevailing doctrine is that of meeting these undeniable disabilities and maladjustments by the philosophy of pragmatic instrumentalism, by a policy of complete "adaptation" to the alleged "individual interests and needs of children," by the abandonment of continuity and of standards, by planless and effortless curricula, and by a general disregard of racial experience.

3. The implications and consequences of this policy have been pointed out. *At many points this philosophy has been shown to be erroneous. It does not cure the disabilities of children. It merely ignores them.* The high school curriculum has become a collection of odds and ends distributed over more than three hundred subjects. Such a system is not only very expensive, but it is essentially futile. It is cruel to the child because it plunges him defenseless into the merciless competition of modern life. It is unjust to society because it leads eventually to incurable problem cases. Above all, *it destroys the common educational background which is indispensable for a democratic society.* At the most critical period in the world's history, when we need social solidarity more than anything else, we have proceeded to establish an educational atomism that is destructive of our finest ideals.

4. *The theorists who advocate unlimited "adaptation" to the child, as well as a multiplicity of standards, have not shown how such a plan can be made to operate in a system of mass education. Logically, it would lead to a special curriculum and to a private tutor for every child.* A few private schools may be able to afford such a luxury. Our tax-supported public schools are unable to carry out such a scheme. Until the money is forthcoming for individual instruction, until classes can be reduced to a maximum of fifteen or twenty, we must continue to depend on the best available group instruction plans. If that is admitted, *children must be trained sequentially and must reach certain reasonable mastery standards in the trunk line subjects.* Otherwise the whole structure is bound to collapse in ruins and utter chaos is inevitable. And that is exactly where we find ourselves at this time.

A planned type of mass education. Shall we say, then, that the only alternative is a return to a rigid system of regimentation such as Russia has again put into operation? Not quite. We shall want to retain all the good features of the "new" education which can consistently be defended and which are not in conflict with common sense and with a reasonable demand for efficiency. In fact, enough experience is already available to show that we may have both mastery of the fundamentals and spontaneous, creative interest on the part of the child. *The supreme demand of the school, now and always, is good teaching and authoritative guidance.*

It is more than a mere theory that *in practically every type of school, and with unselected children, it is possible to prevent the shocking disabilities that are causing so much trouble.* Only, this will never happen by chance. From the moment the child enters the school, his progress must be under close scrutiny. His work must be planned most carefully, from the beginning, with reference to his maturity. At no time must he be allowed to get beyond his depth. And *at all times he must be expected to measure up fully and inexorably to the legitimate tasks he is really capable of doing.* There must be no arbitrary grade goals. A "continuous" progress plan must be substituted. At every level the ideas of "time spent," "units earned," and "ground covered" must be *replaced by the ideal of real understanding and mastery, by excellence of performance.* Under such a plan, disabilities do not arise, or they can be corrected, in the great majority of cases. They are not allowed to become the cumulative menace that is now in evidence everywhere. Is such a plan utopian? Not at all. It is already in operation, as we shall now proceed to show.

Conclusive experiments. In large city schools it is always possible to find cases of retardation dating back several years. A New York City report refers to a certain pupil who remained in the same grade seven terms. Is such a procedure justifiable? "The tuition cost of the terms repeated by the pupils who were enrolled in [New York City] elementary schools in September, 1932, had amounted during the time these pupils had been in school to $21,750,000."[120] For many years that city has been debating plans for eliminating this terrific waste. Twenty-five years ago Dr. Straubenmuller recommended the idea of separating the pupils at the outset into two "streams," an A stream composed of the brighter pupils who are able to meet the usual specifications grade by grade, and a B stream

[120] *High Points,* June, 1935, p. 12.

consisting of those who must advance more slowly and who should not be allowed to develop "a sense of being failures."[121]

The essential features of a "continuous progress" plan, adjusted and perfected in accordance with tested pedagogical and psychological principles, have been in successful operation for some time in a number of educational centers. It has been the writer's good fortune to observe the way this program works, at least in the teaching of arithmetic, in several large public schools experimenting with it. One such school is Public School No. 17 in Rochester, New York, where the plan was introduced and improved by Principal Frank Jenner. Another school of this type is Public School No. 173 in Brooklyn, which until recently had the good fortune of being directed by an outstanding scholar, Dr. S. Badanes. The results were most gratifying in every case. The failure complex was eliminated entirely and the pupils evidently enjoyed their number work. Written reports to that effect were furnished by the teachers concerned. Some of these testimonials read almost like the "before and after taking" advertisements of medicinal firms. It can no longer be doubted that when the primary number work for young children is carried on with a scrupulous regard for their maturity, and when the teaching process stresses understanding and genuine application rather than mechanical drill, there is gratifying retention, and continuous progress becomes a certainty instead of a dubious possibility. This conception of maturation has nothing to do with the mistaken notion that children cannot master the cumulative processes of arithmetic, such as those involving long division, fractions, decimals, and percentage, until they are three or four years older than the customary age. The proposed postponement of arithmetic because of the retardation of so many pupils is not warranted. It would only make a bad situation worse. *When correct methods of instruction are used from the beginning, the mastery of arithmetical processes is a continuous growth phenomenon, each year adding its necessary increment.*

Even in retarded classes that have almost been wrecked by drill devoid of insight, a great transformation is regularly observed when the causes of the difficulty are attacked honestly. Often it is necessary to go back to the very beginning. As a rule, this is the only kind of "remedial" work that really succeeds at any level of instruction.

[121] *Ibid.*, p. 12.

The following illustrative account is based on the written report of an expert remedial teacher whose function it has been for years to adjust the problem cases of a large public school located in a foreign-population district of a typical industrial city. She had soon discovered that no progress whatever was possible without going back to first principles and laying a firm foundation. After that, the work ceased to be hopeless. Such has been her experience from year to year. To quote:

> The case of one class was particularly illuminating. We followed and recorded the work of this group from the fourth grade through the ninth year of the high school. The original analysis showed technical rote memory training. The number *facts* in addition, subtraction, and multiplication seemed to have been assimilated. But the *processes* of subtraction, multiplication, and division were *not understood. There was no evidence of conceptual mastery.* The children admitted that they were "very poor in arithmetic" and had a great dislike for it. Problem solving turned out to be an almost impossible hurdle.
>
> It became necessary to rebuild the entire structure of primary arithmetic for these pupils and to give them the thought training which they lacked. For many months, progress was very slow, necessarily so. No child was any longer permitted to go beyond his depth or to proceed without a conscious feeling of mastery. At last, the class began to move forward at a more rapid rate. *This was not a spontaneous maturation effect, but was due primarily to painstaking daily instruction.* In the sixth grade, these children forged ahead so rapidly that they went beyond the usual prescribed program of work. Most of them are now in ninth grades of the city high schools. A recent follow-up study showed that out of more than thirty pupils of the original group who entered high school, only two were given a "failure" mark in mathematics by their high school teachers. These two pupils had a very peculiar educational history which fully accounted for their difficulties in high school.

In the light of such careful and long-continued experiments we may now assert with confidence (1) *that elementary mathematics is not "too hard" for the average pupil;* (2) *that mental maturity is not attained automatically with advancing years, but is the result of carefully directed mental growth.*

Differentiated Courses and Special Adjustment Programs

The three-track system. When the disabilities of entering pupils are not too marked, it has been found possible in large high schools to apply the principle of selectivity *after* the pupils have been admitted. The well-known *X-Y-Z* system is one illustration of this idea. *It is not an ideal plan.* Because of conflicting schedules, it is rarely possible to organize classes on that basis without a considerable overlap of the various ability levels. That is, a *Y* group will often have many *Z* pupils. Again, *Z* pupils will usually cover far less ground, and even that in a less thorough manner. At the end of a semester or a year, they are usually separated from the achievement level of an *X* or a *Y* group by several months. Naturally, *Z* groups should be given separate sets of achievement tests. Obviously, this involves a special kind of bookkeeping on the official school records. Now, it often happens at the end of a year that a *Z* group cannot be continued intact. In that case, the school usually has no other solution than that of having the subject dropped or putting the remaining *Z* pupils into a normal section. That is, *administratively, the X-Y-Z system usually breaks down, in continuous subjects, after one year. Pedagogically, however, it has considerable merit. While it does not correct the disabilities of retarded pupils, it is less likely to discourage them than is an undifferentiated program.* Finally, a three-track system expects normal and accelerated pupils to rise to a higher level of performance and thus prevents the cult of mediocrity.

The problem of the superior pupil. In our large cities there is a considerable number of brilliant adolescent pupils who are neglected at present. It is from their ranks primarily that we must obtain our future leaders. When their education is held at a level of mediocrity they will either never arrive at the goal they might have reached, or they must make up lost time by an almost crushing burden of work in later years. Physical breakdowns in college may often be traced to an insufficient scholastic foundation. Realizing this situation, many parents have begun to send their children to private schools, often at a great sacrifice. Is it "democratic," then, to neglect the most promising of our pupils in the secondary schools?

The idea of establishing specialized high schools for these outstanding pupils is by no means a new one. In some cities it has been in operation for many years, since certain schools, by common agreement, have been permitted to maintain a very high standard of admission. Such a school is the Brooklyn Technical High School. Another is the time-honored Boston Latin School. The plan has recently been endorsed not only by administrators but also by leading scientists. Assistant Superintendent John L. Tildsley supports it in the following plausible way:

> Of the pupils who took the Terman Test last June, some 8.6% registered I.Q.'s of 120 or over. In our high schools we must have between 20,000 and 25,000 pupils who belong to the class of bright pupils and who are capable of truly intellectual, even creative work. There are, therefore, enough such pupils for two specialized high schools in each of the larger boroughs and one even for Staten Island. For these and for the 175,000 between 90 and 120 I.Q. we should plan an education designed to cause them to realize the full measure of their possibilities in their own interest and that of the State.[122]

Dr. Alexis Carrel recommends this type of selectivity and makes it clear that the selection has nothing to do with class distinctions and similar considerations. He says:

> We must single out the children who are endowed with high potentialities, and develop them as completely as possible. And in this manner give to the nation a non-hereditary aristocracy. Such children may be found in all classes of society, although distinguished men appear more frequently in distinguished families than in others. . . . *It is chiefly through intellectual and moral discipline, and the rejection of the habits of the herd, that we can reconstruct ourselves.* . . . A choice must be made among the multitude of civilized human beings. . . . The only way to obviate the disastrous predominance of the weak is to develop the strong. . . . By making the strong still stronger, we could effectively help the weak.[123]

To prevent wrong impressions, these statements should be read in their context. It will then be seen that they were framed by one of the most enlightened and liberal of our modern thinkers, one who has in mind only an emancipated and truly noble human life.

The problem of the retarded pupil. Until we decide to adopt an integrated mode of foundational instruction such as the one described previously, we may expect to have in our secondary schools, under existing policies, many thousands of very poorly prepared boys and girls. Realizing the seriousness of that situation, the officers of the National Council of the Teachers of Mathematics organized two national committees which have been making a special study of the administrative and pedagogical aspects of the underlying problems. Professor Schorling, of the University of Michigan, has acted as chairman of the committee interested in the techniques of instruction appropriate for retarded pupils. Dr. C. N. Stokes, of

[122] *Ibid.*, p. 14.
[123] Carrel, Alexis, *op. cit.*, [27], pp. 295-297.

Temple University, has investigated the types of classroom organization found feasible by administrators. Both committees have repeatedly submitted preliminary reports of their findings, which were published in various issues of *The Mathematics Teacher*.[124]

Particular attention should be called to a supplementary monograph prepared by Professor Schorling, entitled *The Technique of Instruction for Dull Normal Pupils*, issued by the Bureau of Educational Reference and Research, Ann Arbor, Michigan. It contains valuable suggestions and findings, important excerpts and summaries from recent studies, as well as an extensive bibliography. Of 144 pupils classified as dull and enrolled in the seventh grades of a Michigan school system, only 3.5 per cent had a reading ability corresponding to the seventh-grade level, and only 6.9 per cent had reached the seventh-grade norm in arithmetic. As many as 29.9 per cent were only at a fourth-year level in reading, and 20.8 per cent had not gone beyond the norm of that grade in arithmetic. One of the most significant contributions of this study is the statement that *"the curve of growth toward a specific maturation under constant environmental influences appears to be the same for the dull as for the bright."*

The best available plan for retarded pupils entering the seventh grade is that of giving these pupils a special two-year training course intended to remedy their weaknesses, thus building up the necessary maturity for high school courses at the minimum level. Such a course has been tried in a number of school systems.[125] It should include work in English, basic mathematics, social studies, and modern arts and crafts, much attention being given to health factors and to adequate recreation. The best results are obtained when the teachers of these pupils have had grammar school experience, since high school teachers rarely know the program of the elementary grades. At the end of such a two-year course the retarded pupil is ready for a reduced eighth-grade program, after which he may continue in school with a more reasonable expectation of success.

[124] *The Mathematics Teacher,* November, 1932, pp. 420-426; October, 1933, pp. 350-365.

[125] See Feingold, G. A., "Intelligence and Persistency in High School Attendance," School and Society, October 13, 1923. See also the same author's article on "The Basic Function of Secondary Education," *op. cit.,* [51]; and *High Points,* June, 1935, [96], p. 14.

PART SIX

THE TEACHING PROCESS AND THE TRAINING OF TEACHERS

The Impending Revision of Educational Psychology

Conflicting theories of learning and teaching. We have seen how profoundly our schoolroom procedures have been affected not only by the dominant philosophic ideas of our period, but also by current educational doctrines. It is now necessary to show how the teaching process itself is being affected by changes in *psychological* theory. Gradually, it is being realized that a whole generation of teachers have been misled by mechanistic laws of learning. When this fact is finally understood, we may expect a great transformation in classroom methods, in the construction of textbooks, and in our ordinary testing machinery. At present, three conflicting theories of learning are competing with each other. They have been described with scholarly clearness by Dr. William H. Brownell of Duke University in the *Tenth Yearbook of the National Council of Teachers of Mathematics.*[126] We submit a brief exposition of these instructional theories.

"Bond" theory of learning. Under the stimulus-response hypothesis of learning, it is claimed that "learning is connecting." Whenever there is a response of the learner to a stimulus, a "bond" is formed. This "bond" is conceived as involving changes in the nervous tissue. It is believed by advocates of the "bond theory" that the strength and the permanence of a "bond" are largely a matter of repetition, of "exercise." That is, the "trace" left in the nervous organism by each response is strengthened by repetition, by drill. Moreover, it is claimed that the learner is more likely to make progress if the "bonds" are established in a pleasant way, if the "effect" on the learner is agreeable rather than painful. The more attractive or pleasurable the learning process is made, the more lasting its results are going to be. The "Law of Exercise" and the "Law of Effect" may be regarded as the backbone of the "bond theory" of learning.

[126] Brownell, W. H., "Psychological Considerations in the Learning and the Teaching of Arithmetic," in *The Tenth Yearbook of the National Council of Teachers of Mathematics,* pp. 1-31, Bureau of Publications, Teachers College, Columbia University, New York, 1935.

If this theory is held to be correct, it becomes necessary to ascertain by a careful analysis of each subject-matter field precisely what are the "bonds" to be established, and how much *repetition* or drill is necessary to make them permanent.

Drill theory in arithmetic. The appalling consequences of such a mechanistic conception of the learning process have been demonstrated with dramatic force in the field of arithmetic. The drill conception of arithmetic has been outlined as follows by Dr. Brownell:

> Arithmetic consists of a vast host of unrelated facts and relatively independent skills. The pupil acquires the facts by repeating them over and over again until he is able to recall them immediately and correctly. He develops the skills by going through the processes in question until he can perform the required operations automatically and accurately. The teacher need give little time to instructing the pupil in the meaning of what he is learning: the ideas and skills involved are either so simple as to be obvious even to the beginner, or else they are so abstruse as to suggest the postponement of explanations until the child is older and is better able to grasp their meaning. The main points in the theory are: (1) arithmetic, for the purposes of learning and teaching, may be analyzed into a great many units or elements of knowledge and skill which are comparatively separate and unconnected; (2) the pupil is to master these almost innumerable elements whether he understands them or not; (3) the pupil is to learn these elements in the form in which he will subsequently use them; and (4) the pupil will attain these ends most economically and most completely through formal repetition.[127]

Under the impact of the drill theory, courses of study and textbooks have been forced to allot an ever-increasing amount of time and space to mechanical repetition. And it can hardly be denied that the drill theory at present is by far the most popular method of instruction.

> In the classroom its popularity is manifest in the common extreme reliance upon flash cards and other types of rapid drill exercises, in the widespread use of workbooks and other forms of unsupervised practice, and in the greater concern of the teacher with the pupil's speedy computation and correct answer than with the processes which lead to that computation and that answer.
>
> But the popularity of the drill theory is by no means revealed only by the prevalence of certain practices in classroom instruction. On the contrary, its popularity is evident, as well, in the organization of arithmetic textbooks, in much of the research in arithmetic, in current

[127] *Ibid.*, p. 2.

practices in measuring achievement in arithmetic, and in treatises on the teaching of arithmetic.[128]

Perhaps this overwhelming prevalence of mechanical drill is due not merely to its supposed scientific soundness. In even larger measure it may be caused by our preference for mechanical ways of solving all the major problems of our day. That is, the machine age views the mind as a mental machine capable only of mechanical reactions.

Absurdities of the drill theory. It is not possible, because of a limitation of space, to discuss in detail the flaws of the "bond" theory. There is increasing agreement, however, that it is essentially false. In particular, the Law of Exercise has been restated by Thorndike himself and has been entirely discarded by other psychologists. A similar statement holds with reference to the Law of Effect.[129]

There is little doubt that the drill theory has almost ruined the teaching of arithmetic. When a teacher is a mere drill sergeant who pays no attention to number concepts, to underlying meanings and principles, we may expect exactly what we find today—wholesale failure in arithmetic. *Here is the deepest cause of the present breakdown of mathematics in the elementary and the secondary school.*

The fatal weakness of the drill theory is that it ignores the all-important factor of meaning and that it confuses mechanical reaction with thinking. Above all, *"the drill theory sets for the child*

[128] *Ibid.*, p. 3.

[129] For more detailed accounts of present trends in educational psychology, including the problem of drill and repetition, see, for example, Bode, Boyd, *Conflicting Psychologies of Learning,* D. C. Heath and Co., Boston, 1929; Brown and Feder, "Thorndike's Theory of Learning as Gestalt Psychology," *Psychological Bulletin,* 1934, *31,* pp. 426-437; Cason, H., "Criticism of the Laws of Exercise and Effect," *Psychological Review,* 1924, *31,* pp. 397-417; Dunlap, Knight, *Habits,* Liveright, Inc., New York, 1932; Hartman, George W., *Gestalt Psychology,* The Ronald Press Co., New York, 1935; Heidbreder, Edna, *Seven Psychologies,* The Century Co., New York, 1933; Jastrow, J., "The Current Chaos in Psychology," *Scientific Monthly,* August, 1935, pp. 97-110; Jastrow, J., "Has Psychology Failed?", *The American Scholar,* Vol. 4, No. 3, 1935, pp. 261-269; Murchison, C. (Editor), *Psychologies of 1930,* Clark University Press, Worcester, 1930; Murphy, Gardner, *General Psychology,* Harper & Brothers, New York, 1933; Pintner et al., *An Outline of Educational Psychology,* Barnes and Noble, Inc., New York, 1934; Thorndike, E. L., *Human Learning,* The Century Co., New York, 1931; Wheeler, R. H., *The Laws of Human Nature,* D. Appleton Co., New York, 1932; Wheeler, R. H., *The Science of Psychology,* Thomas Y. Crowell Co., New York, 1929; and Wheeler, R. H. and Perkins, F. T., *Principles of Mental Development,* Thomas Y. Crowell Co., New York, 1932.

a learning task the magnitude of which predetermines him to failure."[130] Thus, in the field of combinations involving merely whole numbers, more than 1,600 number facts are to be learned, which the child should know at the end of his fourth year in school.

Similar analyses of other phases of arithmetic have resulted in item-totals equally staggering. Thus, Knight demonstrates the presence of 55 "unit skills" in the one process of division of common fractions, and Brueckner finds 53 "types" of examples (exclusive of "freak types") in the subtraction of common fractions alone. In problem-solving Judd reports, on the basis of an examination of only three sets of textbooks, approximately 1,900 different ways of expressing the fundamental operations in one-step verbal problems, and Monroe and Clark are able to differentiate 333 kinds of verbal problems (52 kinds of "operative problems" and 281 kinds of "activity problems") with which children must eventually be able to deal. These figures, large as they are, can be regarded only as typical of what would be found if the many other aspects of arithmetic were dissected as have been the relatively few reviewed above.

The statement that the drill theory in its extreme form sets an impossible learning task for the child would seem to be justified by the results of the analyses mentioned above.[131]

The story becomes even more complicated when we add the doctrine that the various number "facts" must be arranged and learned in the order of their relative difficulty. Thus, it is claimed that certain combinations have been shown statistically to be many times as hard as others. Hence we have "easy" and "hard" number facts.

Just what value these figures [of statistical degrees of difficulty] have is problematical. Have they, for example, been collected from a fair sample? Why should the order of frequency of error be a sign of the degree of difficulty? And what, exactly, does "difficult" mean in relation to a mass of unrelated facts? Is difficulty of understanding or difficulty of remembering the source of error? And, to go at once to the root of the matter, is the occurrence of error and in particular of such bizarre combinations of errors a sign of the incapacity of the child or of the inadequacy of the teaching?

That a statistician trained in the interpretation of his results and conscious of the limitations of his method would reach the second of these conclusions, I am personally convinced. And my conviction is all the more strengthened by a study of the volumes on the teaching of early number, of which the books already cited are but the better

[130] *Tenth Yearbook of the National Council of Teachers of Mathematics,* [126], p. 6.

[131] *Ibid.*, p. 7.

examples. *In all their pages, there is no word, not one, in regard to the meaning of mathematical processes. Consequently there is no way whatever within an intelligent child's reach for correcting his own errors in the light of a well understood process. In these books, there are no processes; there are only statistical facts. . . .*

In short, *the statistical method as applied to elementary mathematics has had the sublime result of eliminating mathematics from the curriculum. As matters now stand, arithmetic is merely a device for training errand boys to come back from the grocer's with the right change.*[132]

It is now known that the hierarchy of difficulty imposed upon the elementary number facts is likewise in need of reconsideration. Mr. L. C. Thiele carried on experiments in the Detroit Public Schools in order to find out the relative merits of the drill theory and of generalized methods based on concepts and on insight. Mr. Thiele writes:

The pupils who employed the generalized method learned the combinations of larger numbers almost as well as those of the smaller numbers. For example, $7 + 9$ and $9 + 7$, which are listed as the most difficult of all the addition combinations on Clapp's list, were not any more difficult than $5 + 2$, $4 + 2$, and $9 + 3$ for the pupils who were studying according to the generalized method.[133]

Incidental learning theory. In their justified abhorrence of purely mechanical drill methods, "progressive" teachers looked for a substitute. They began to feel that children will perhaps "learn as much arithmetic as they need, and will learn it better, if they are *not* systematically taught arithmetic," but establish their contact with that subject through such natural quantitative situations as arise in the various activities of the classroom. That is, the number facts were to be acquired through *incidental experience.*

This "incidental learning theory" has now been tested thoroughly by groups of teachers who believe in the activity idea. Perhaps the most authoritative study of this sort ever undertaken in the field of arithmetic was that sponsored by Dr. Paul R. Hanna, formerly of Lincoln School of Teachers College, Columbia University. Fourteen experienced teachers from private and public schools participated in the study. Its purpose was that of investigating "oppor-

[132] DeGrange, M., "Statisticians, Dull Children, and Psychologists," *Educational Administration and Supervision,* November, 1931, p. 565.

[133] Thiele, C. L., "The Mathematical Viewpoint Applied to the Teaching of Elementary School Arithmetic," *The Tenth Yearbook of the National Council of Teachers Mathematics,* [126], p. 221.

tunities for the use of arithmetic in an activity program."[134] The summary of findings given at the end of the published report describing this investigation leaves no room for doubt that "incidental learning" will not produce results in arithmetic. *We may regard it as an established fact that while activities and projects may prove extremely valuable and indispensable as a means of motivating arithmetic, they will not automatically lead to an understanding of its concepts, principles, and processes.*

Theory of conceptual and meaningful teaching. According to this theory it is of decisive importance whether or not children *understand* what they learn. If they see no sense in what they are expected to do, no amount of repetition or mechanical drill is likely to secure retention or correct application. A parrot may repeat a phrase countless times without arriving at a rational comprehension of it.

Recent psychological thought is virtually committed to the idea that *mere repetition does not guarantee learning*, and that *the learning response rather than the learned response is of primary importance.*[135] Professor Wheeler has stated the futility of mere repetition very strikingly in this manner:

> Drop a ball a million times and it will not fall more easily the last time than the first. . . . An electric current will travel just as well through a switch the first time as it will travel the thousandth. Repetition makes no difference to the switch, nor to the current. . . . To assert that we learn by doing, or learn through exercise, is as meaningless as to claim that we grow by living in time.[136]

According to Professor Herrick, "*some stupendous feats of learning can apparently be done in the twinkling of an eye,* as when one 'sees through' a difficult situation in a flash of understanding."[137]

This is by no means a new idea. The central thought was stressed as early as 1910 by Professor Dewey, when he said, "*Practical skill, modes of effective technique, can be intelligently, non-mechanically used, only when intelligence has played a part in their acquisition.*"[138]

[134] Hanna, P. R., "Opportunities for the Use of Arithmetic in an Activity Program," *The Tenth Yearbook of the National Council of Teachers of Mathematics,* [126], pp. 85-120.

[135] Dunlap, Knight, *op. cit.*, [129], pp. 33 ff.

[136] Wheeler, R. H., *The Science of Psychology,* [129], p. 323.

[137] Herrick, C. J., *The Thinking Machine,* p. 241, The University of Chicago Press, Chicago, 1929.

[138] Dewey, John, *How We Think,* p. 52, D. C. Heath and Co., New York, 1910.

As to mathematics, we are realizing at last that it is *"a closely knit system of understandable ideas, principles, and processes." Mathematics is based on concepts.* The mastery of these basic concepts on the part of the learner is the most essential problem of the teacher. When we take this view of the teaching of mathematics, we cease to have any use for the idea of confronting the pupil with a "heterogeneous mass of unrelated elements" to be mastered by repetition. Instead, *the emphasis must be on a real comprehension of mathematical principles and relationships.*

Now, the development and mastery of clear concepts takes time. It should never be hurried. *The generalizations of any subject cannot be acquired in a few lessons.* The fundamental discoveries in any field cannot be made at one stroke; the learner must be confronted with ever new situations demanding a repetition of these successive discoveries, and not a premature memorization of facts.

It follows that the initial stages of each learning process must be gauged carefully with reference to the maturity of the learner. Important ideas and processes must be "spread" over a wider space of time than is usually the case. "Spaced" learning is more effective than "bunched" learning.[139]

[139] This fact alone should raise grave doubts as to the soundness of the Manchester, New Hampshire, experiment, recently reported so confidently in three successive issues of the *Journal of the National Education Association* (November, 1935 to January, 1936). Under this plan, formal work in arithmetic is virtually abandoned in the first five grades, and intensive work is then begun in the sixth grade. The new program, as announced by Superintendent L. P. Benezet, lays commendable stress on reasoning and on the avoidance of purely mechanical drill. It is urged that children be "made to understand the reason for the processes which they use." That is as it should be. Nevertheless, such are the grave defects of the new curriculum as a whole that the ultimate consequences are bound to be disappointing, if not disastrous. Only a few of these can be mentioned at this point. (1) Mr. Benezet reports that splendid results accompanied even this moderate exposure to arithmetic, as soon as it was taught on a rational basis. How can any one doubt, therefore, that even better results would be obtained if *real thinking* and *genuine motivation* characterized the study of arithmetic *from the beginning*, and in a curriculum of normal length? (2) A period of only 20 or 25 minutes a day is set aside for arithmetic in grade VI, of 25 minutes in grade VII-B, and of 30 minutes in grade VII-A and in grade VIII. The attempt to cover such a large number of concepts, principles, and processes within such a totally inadequate time schedule must inevitably lead to imitative repetition, or to the abandonment of real mastery, thus vitiating the whole plan. Mr. Benezet states that abundant work in *mental arithmetic*, involving the solution of problems, "is far more important than accuracy in the four fundamental processes." One wonders whether such solutions, oral or written, are of any value if the answers are *not* accurate. (3) The Manchester plan would certainly make impossible the enlarged mathematical program which schools of the junior high school type have adopted so largely in recent

Some of the main principles of the new psychology of learning have been outlined by Professor Wheeler, one of its leading exponents, as follows:

1. *Learning is a function of maturation and insight.* It is a growth process that follows laws of dynamics, that is, laws of structured, unitary, energy systems or fields.

2. First impressions are of total situations, but are undifferentiated. First movements are mass actions, likewise undifferentiated. In spite of appearances to the contrary, these impressions and movements are completely integrated. . . .

3. *Learning is not exclusively an inductive process.* First impressions are not chaotic and unorganized. They are merely unstable in the sense of not being under environmental control. . . .

Learning, then, is not a matter of forming bonds, a process of putting pieces of experience together. *It is not based on drill and on repetition of response. Bond psychology is irrational and has never been required by the facts of observation.* It is a mechanistic philosophy imposed upon the facts. On the contrary, *learning is a logical process and from the beginning characterized by a grasp of relations, no matter how vague. Progress is systematic; it is a logical expansion and differentiation of unitary grasps of total situations—of wholes. It is organized and insightful, creative response to stimulus-patterns.*

4. *Learning does not proceed by trial and error. This concept is based upon an illusion, the fallacy of the double standard,* arising out of the difference between the adult and the animal or child. *There is no such thing as a trial-and-error process anywhere in nature. . . .*

5. *More important, by far, than formal, prescribed methods of instruction are the personality of the learner and of the teacher, and the relationship between these personalities.* Learning is subordinate to the growth and the demands of the personality-as-a-whole. *The atmosphere of the classroom is more important than textbooks.* The latter are necessary, but are secondary.

6. *Learning depends upon the will to learn, which cannot be forced*

---

years, after such strenuous and long-continued efforts on behalf of mathematical reform in this country. In other words, the splendid progress achieved at last during the past generation is to be swept away by this latest panacea. Is it possible to question the ominous effects of Mr. Benezet's curtailed curriculum on the mathematical preparation of at least three million pupils? Why should American children again be placed at such a disadvantage, as compared with the adolescent pupils of all other leading nations?

By way of a further refutation of the Manchester plan, see Buckingham, B. R., "When to Begin the Teaching of Arithmetic," *Childhood Education,* May, 1935; MacLatchey, Josephine, "Number Abilities of First Grade Children," *Childhood Education,* May, 1935; also Buckingham, B. R. and MacLatchey, J., "The Number Abilities of Children When They Enter Grade One," *The Twenty-ninth Yearbook of the National Society for the Study of Education,* pp. 473-524, Public School Publishing Co., Bloomington, Ill., 1930.

by requirements or authority, but must be challenged by dynamic teachers and dynamic teaching.

7. *Learning depends on clearness of goals,* and the fitness with which tasks are adjusted to the pupil's level of maturation and insight. *Progress is made by pacing.*

8. Goals are their own rewards, under natural law. Grades, grade points, many forms of motivation by social competition, and other hypocrisies are detrimental to learning. *The subject must be worth learning in its own right. It can be made so, easily.*[140]

Transfer of training. We have repeatedly referred to the current misunderstanding of the problem of generalized training, incorrectly identified with "mental discipline." The history of the controversy relating to formal discipline has been made available in a number of sources.[141] In its modern setting, under the name of the "transfer of training," the doctrine is anything but "obsolete." Particular attention is again directed to the recent foundational work of Professor Lashley and of Dr. Orata. Their studies should be made available to every teacher.

The key sentence of the modern doctrine may be stated as follows: *"From the standpoint of the teacher and the school, the solution of the problem of transfer of training is to train for transfer."*[142] In other words, transfer is not an automatic phenomenon. It is, as Orata happily phrases it, "a technological problem."

The new theory of transfer and of conceptual learning throws a clear light on the all-too-prevalent disabilities of school children and high school pupils. Not having been taught foundationally, being without a real understanding of concepts and principles, they are unable either to remember or to apply the skills they seemed to know so well. We have thus made it virtually impossible for the pupil to succeed.

Necessary changes in teaching procedures. Enough has been said about the new psychology of learning to make it obvious that *we must turn away from mechanical teaching and must emphasize meaning, insight, intelligence.* In no other field is this change so

[140] Wheeler, R. H., "The New Psychology of Learning," *The Tenth Yearbook of the National Council of Teachers of Mathematics,* [126], pp. 237-239. See also pp. 19-31 of article by W. A. Brownell in the same volume, [126].

[141] See, for example, Betz, W., "The Transfer of Training," *The Fifth Yearbook of the National Council of Teachers of Mathematics,* pp. 149-198, Bureau of Publications, Teachers College, Columbia University, New York, 1930.

[142] Orata, P. T., *op. cit.*, [92], p. 269; Lashley, K., *op. cit.*, [92], especially pp. 172-173.

imperative, and yet so easily made, as in mathematics. From the first lesson in arithmetic to the last lessons in higher mathematics the fundamental issue is the same. Hence we shall close these remarks on the new psychology with the following illuminating passages which, while relating to arithmetic, stress the continuous importance of meaningful instruction and may be applied generally. Thus Dr. Brownell summarizes the importance of the "meaning" theory in this manner:

Most of the illustrations of, and arguments for, the "meaning" theory have been drawn from the field of primary number. This fact should not, however, be interpreted to mean that the theory holds only for the first three grades. On the contrary, *meaning affords the soundest foundation for arithmetical learning throughout the elementary school.* Primary number has been most often cited for another reason. Almost everyone agrees that children in Grades 5, 6, 7, and 8 have to be able to "think" in arithmetic. That ability to "think" in these grades is conditioned by "thinking" in the primary grades is a fact which is much less commonly recognized. *No one has shown how it is possible for children suddenly to become intelligent in upper-grade arithmetic when they have been allowed no exercise of intellignce in lower-grade arithmetic.* In spite of the unreasonableness of such an expectation, primary number is taught as if skills acquired mechanically would later surely take on meaning, and verbalizations memorized unintelligently would later inevitably become well-rounded concepts. *It is the thesis of the "meaning" theory that children must from the start see arithmetic as an intelligible system if they are ever to be intelligent in arithmetic.*[143]

In his *Psychological Analysis of the Fundamentals of Arithmetic*, Professor Judd wrote:

The issue between *two fundamentally different views* with regard to the curriculum which are now before the school people of this country is nowhere clearer than in the sphere of number training. *It is the issue between the view that the duty of the school is the cultivation of comprehensive general ideas and the view that the sole duty of the school is to train pupils in relatively trivial particular skills. There can be no doubt that powerful influences are at work in the pedagogical world to reduce all training to the cultivation of practical and particular skills.* Fortunately, the mind of man is so organized that it generalizes. *Even if all the curriculum-makers resolve to train nothing but particular abilities, pupils will generalize and will continue to do what the race has done throughout its history, that is, abstract from particular situations those aspects which are most universal.* Some

[143] *The Tenth Yearbook of the National Council of Teachers of Mathematics*, [126], p. 31.

children will acquire the general idea of mathematical exactness no matter how far curriculum-makers go in running counter to human history.

All the experiments and analyses reported in this monograph lead to the conclusion that *general ideas are the most important products of instruction in arithmetic. The fundamentals of arithmetic are general ideas and general formulas, not a multitude of special skills.*[144]

Let us look forward to the day when the convincing and tested principles outlined above will be followed generally in every classroom and at each level of instruction. Toward the arrival of that day we must bend all our energies.

The Teacher in the School of Tomorrow

Present status of the teacher. The majority of teachers have always been anxious, as have been the members of other leading professions, to render the best service of which they are capable. That they often fall short of the ideal qualifications which modern education has suggested for those to be entrusted with the great responsibility of preparing the youth of the land for effective citizenship and for a worthwhile life, is only too apparent. It is hardly necessary to elaborate this point in detail. Few secondary teachers at present bring to their teaching career either a comprehensive scholarship or an adequate pedagogic equipment. A very large number have "taken" the prescribed college course, and have been duly credited with the requisite number of "points" and "hours." Their cultural background, however, is either too narrowly specialized or too superficial to command general respect. Whose fault is it that we are struggling with such serious professional handicaps?

Professor David Eugene Smith, writing on these problems from the serene pinnacle of a lifetime rich in service and in scholarly pursuits, ponders some of the reforms that seem to be imperative.

If culture courses for all teachers were established, who would administer them? There is the difficulty and it is a very great one. The man must be an inspiring teacher, filled with the enthusiasm which the subject engenders, and one of broad knowledge of those things which concern the finer instincts of mankind. He must be able to appeal to all whose tastes are not hopelessly distorted by the radio programs, the motion pictures, the cocktail party, or the bridge table. *He should be the best teacher in the school of education—best in his*

[144] Judd, C. H., *The Psychological Analysis of the Fundamentals of Arithmetic*, pp. 116-117, The University of Chicago Press, Chicago, 1927.

range of knowledge, best in his balanced mind, best in his sympathy with those things of life which weigh heavier than wealth, and with which one may spend his time happily in the home as well as in the palaces of art or in the great libraries of the world.[145]

That there has been a decline from the high regard which teachers enjoyed not so many years ago is brought out by the same writer in the following passage:

In approaching this phase of the matter I confess to a feeling of helplessness when I compare the position in society held by the old-time schoolmaster with that of teachers in high schools and colleges of today. In my youth the teacher in the "academy" was looked up to with respect as a learned man. He dressed the part, acted the part, and in general honored it. Today a teacher in an equivalent position in a high school holds, in the opinion of his fellow citizens, a much lower rank. He plays bridge better, but he is rarely appealed to in a matter of public importance; and any topic relating to literature, art, or religion seems much more foreign to his interests than it was to his professional ancestors. In short, the schoolmaster of a half-century ago was looked upon as a man of superior intelligence, whereas today the average school-teacher has lost that rank. One of two causes may explain this situation, either the teacher knows relatively less or the community knows relatively more. We are led therefore to ask if the mass of citizens know relatively more than they did fifty years ago, and whether their tastes are better than was formerly the case. For the same reason *we are led to ask if the teachers are really as well qualified, not in the science of education but in their general culture.* Are parents more often sending their children to private schools because of the poorer teaching in the public ones?[146]

And why do teachers not succeed more frequently in overcoming the defects of their original preparation? Or why do they not strive to raise to a much higher level the excellence of their professional work? Professor Smith prefers not to answer these questions with finality, but submits searching queries that suggest his own position. He says:

All this leads us to a very serious question—is the instruction less efficient because the teachers in our city systems have no longer the time for a normal and cultural way of living? Burdened as they are at present by seemingly endless reports, by useless and wearisome committee meetings, and by other activities which sap their energies, how can they do their best work in making the hours passed in school

[145] Smith, D. E., "Certain Immediate Problems Confronting American Schools of Education," *Teachers College Record*, March, 1935, p. 472. See also Holmes, Henry W., "Pedagogical Racketeering," *Scribner's Magazine*, March, 1935.

[146] Smith, *op. cit.*, pp. 475-476.

a period of pleasure as well as profit for the pupils? Is not the drudgery of reading masses of papers outside the schoolroom one of the reasons why young people of promise are coming to look upon teaching as too much like the task of a day laborer? Is not this the reason why they are seeking other means of living than what should be one of the noblest and most interesting professions they could choose?

The bearing of this phase of our discussion lies right here: Where, after all, lies the blame for this overburdening of teachers with nerve-racking work outside of school hours? Is there no way of closing these educational sweatshops? Where, I repeat, lie the blame and the remedy? Where were these teachers taught to examine with wearied eyes the results of endless "tests" or to compute correlation coefficients? If these activities are necessary for the school, why are they not performed by paid assistants with calculating machines to help when needed? *Where lies the blame if not in schools of education in which all this work is recommended to principals and superintendents?* And, furthermore, is it recommended by professors who themselves spend their extramural hours in doing this kind of work in connection with their own classes? And, still further, do the principals and superintendents likewise spend their evenings in the same way?

I do not know the answers to those questions, but I feel that we are all interested in helping to solve such problems.[147]

A challenge to teachers. Two further demands, in addition to those of *general culture* and of *adequate professional training*, are now admitted to take precedence over all others. The first is to the effect that all teachers must have *absolute mastery of the subject-matter which they desire to teach.* The second has to do with the subtle but vastly important factor of the *personality* of the teacher.

It is doubtful whether any one has stated the issue of the master teacher with greater force than have the noted authors of the psychological treatise entitled *Principles of Mental Development*:

It is the personality of the teacher by means of which the evil effects of mass education must be reduced. We have seen that *learning is directed growth*, the expansion of a personality that controls the process of learning from within. But it is growth of a human being dependent upon motivation through personality contact and the culture which the personality in the teacher derives from the culture of the race. There are teachers who brighten the schoolroom with their personality and enthusiasm, and who inspire the pupil with the romantic and exciting color of the information which they divulge. They know life, and live it. They see and understand discovery and creation. *They are the doorways into a wondrous world. It is from these that the pupil learns.* There are teachers whose rooms are as gloomy as a

[147] *Ibid.*, pp. 476-477.

tomb; teachers who rock the atmosphere with the cyclones of an erratic temperament. There are others whose indifference chills and subdues spontaneity; those whose superciliousness makes the pupils stiffen with bridled protest and contempt; those against whose fawning manner even a child revolts; and those whose sheer emptiness casts a pall upon the schoolroom. . . . *No student passes through the educational system without the stamp of some teacher upon him.* Thousands are inspired to careers of particular achievement by their teachers, alone; thousands lose the benefits of school because of treatment from their teachers. Unnumbered mental lives suffer septic operations and endure the ravages of permanent infection, through sheer inefficiency and stupidity of teaching. *The teacher is the physician of maturative processes, and there is no redress from wounds that are sewed up wrong.*

Grades, hours of credit, these are most meticulously inspected, but the sterilized insight, the unfulfilled demands and the disillusionment, who knows about them? Not the system. It is not organized to care. Personalities retard mental growth; they raise intelligence as measured by test scores from the lower portions of a given range to the highest. It is being done every day. It is the most pressing need of present Education, a more adequate training and selection of teachers. And why are not more of the cream of college youth, with the personalities to undertake this vital work, taught of its importance, and shown the opportunity? Ask almost any better student of today if he is planning to enter Education and his reply is, "I should say not!" The educational dryness, dullness, mechanical deadness of the mill, so fresh in his mind from just having passed through it, is too much. The intelligence ratings of a Teachers College population are not encouraging! The financial outlook for the teacher is anything but tempting to the individual with ambition and personality. . . . When a system puts through hundreds of thousands of individuals, most of them shouting with gladness to have left it, something is wrong. It is not living. Achievement leaves a memory mixed with pride and joy; the feeling of something unfinished that demands a further completion. Learning is, by nature, achievement; but where is the feeling of something unfinished? . . . Why the many thousands who go on from high school with no demand to reach a real intellectual goal or to prepare themselves, seriously, for the task of solving the major problems of the race?[148]

The teacher as the engineer of human nature. Intelligent critics of our age have told us that thus far our civilization has rested too exclusively on the study and the conquest of external nature. We have codified and glorified the laws of matter and have assigned a place of supreme importance to the *mechanical* engineer. We are ready for a change. "The day has come to begin the work of our

[148] Wheeler, R. H. and Perkins, F. T., *op. cit.*, [129], pp. 509-511.

renovation. . . . We must liberate ourselves from blind technology and must grasp the complexity and the wealth of our own nature." We are beginning to understand "that man is not a machine, that the laws of his behavior are laws of intelligence, will and personality, not the laws of association."[149]

The teacher of tomorrow will be a *human* engineer. He will realize that "teaching is not a mechanical process of telling facts. It is a contact made between a personality and a group of personalities, in which the imparting of knowledge is subordinate to the vitality of meaning which the teacher gives the facts by his attitudes, interest in the subject, breadth of grasp of the subject, and by the atmosphere of the classroom which he engenders."[150]

In the school of tomorrow the teacher will regain the esteem he once enjoyed, because he will be an artist, a scholar, and a friend. He will direct the activities of the classroom with consummate skill, stimulating and effecting desirable mental, moral, and spiritual growth. And the community of tomorrow will reward him by granting him the freedom of his great profession, economic security, and a position of confidence and trusted leadership.

CONCLUSIONS

General summary. The salient characteristics of educational theories and practices, which have been reviewed in this discussion, may be briefly generalized as indicated in the following statements.

1. We have seen that education is facing a world in turmoil, that the machine age and technology have led to a profound transformation of society and its institutions, and that mechanistic science has made us uncertain of values and goals. We have lost our perspective and the unity of our souls.

The present crisis in education was shown to be essentially an outgrowth of this crisis in human affairs. Above all, we must recapture a sense of the wholeness of culture, and a regard for the unity of knowledge and the integrity of the social order. To the mechanization of modern life must be opposed the unique contributions of the individual. We must strive for a unified grasp of total situations and abandon our atomistic orientation.

2. It is characteristic of this period that our best educational

[149] *Ibid.*, p. ix.
[150] *Ibid.*, p. x.

efforts are being neutralized by a lack of vision that is symptomatic of our conflicting educational philosophies. The dominant philosophy of this generation is that of pragmatic instrumentalism. Its grave flaws were pointed out. Educational sanity demands a rejection of these erroneous positions. The philosophy of change and of social reconstruction was contrasted with the concept of invariance and the doctrine of permanent values. There can be no educational recovery until the idea of "change at any cost" is replaced by a comprehensive and consistent philosophy resting on a solid foundation of ultimate objectives.

3. The present status of "progressive" education was examined in detail. The characteristics of the activity program were outlined and the serious weaknesses of activism were pointed out. Our analysis of "progressive" education revealed some fatal "danger zones" that must be eliminated if the influence of this movement is to be considered either beneficial or lasting.

4. It was shown that for many years the curriculum has been a source of confusion. The various stages of curriculum revision during the past four decades were reviewed. Four new ways of dealing with the curriculum were summarized and criticized. The continual "adjustment" of the curriculum has made it a collection of odds and ends, for which no one can have respect. It was urged that, instead, the curriculum be built on enduring backgrounds and on basic frames of reference. Six trunk-line domains were suggested as its permanent categories. The importance of continuity and of an adequate time allowance was stressed. The current "unit" system stands in the way of a desirable integration and must be given up. Finally, the unique importance of mathematics was discussed, over against the constant attacks that are being made on mathematical teaching.

5. Consideration was given to the crucial problem of mass education. The measurement movement revealed a disturbing amount of retardation and maladjustment at all stages of the educational process. Erroneous assumptions and inferences have led to a widespread indorsement of the policy of unlimited "adaptation" to children's individual "needs and aptitudes." The various attempts at adaptation were described. They have failed to produce satisfactory results, thus proving that the fundamental causes of the difficulties to be corrected have not been met. In particular, the notion that vocationalism is a cure-all for educational retardation was

attacked as utterly fallacious, and due attention was given to authoritative recent views as to the future of vocational training.

The ominous effect of "adaptation" on standards was considered at length. Evidence was submitted that we must return to a defense of legitimate and binding educational norms. It was maintained that the vexing problems of mass education will not be solved by ignoring them. Only a planned type of mass education will be of avail. This involves an integrated program of foundational instruction based on the idea of continuous progress made possible by genuine understanding and real mastery. Within this general setting, attention was also given to the feasibility of specialized high schools for superior pupils and of adjustment courses for retarded pupils.

6. In the light of the new psychology of learning the teaching process was shown to be in need of revision. The drill theory of teaching was held to be one of the most obvious causes of wholesale educational failure. It must be replaced by a type of instruction that is based on conceptual and meaningful learning, real motivation and the genuine co-operation of the learner. The idea of "training for transfer" was seen to be the modern and truly scientific successor to the doctrine of formal discipline.

The present status of the teacher was described as unsatisfactory. The teacher of tomorrow will not be a mere technician, but will be trained more effectively along broadly cultural lines. In the community of tomorrow this master teacher will regain a commanding position, commensurate with his potential qualities of real leadership.

The outlook. During the decades that lie ahead, secondary education will assume an even greater importance than in the past. The secondary school will become the people's university. But it can discharge that lofty mission only if its basic philosophy is sound. It must avoid fantastic educational theories and must be built on the bed-rock of permanent values, and on the ideal of social solidarity. The curricula of the emerging high school must be integrated around the enduring trunk-line categories that date back to the dawn of history. There must be no opportunist tinkering with values and standards. Good teaching and efficient guidance must be in evidence in every classroom.

America's cultural mission will depend in large measure on what happens in the American high school. If we are satisfied with mediocre attainments, our national life will be characterized by mediocrity. If, on the other hand, our young people are at all times imbued with ideals of excellence, the lofty visions of the founding fathers will be increasingly realized.[151] Conceived in this way, the secondary school will become, as Dr. Abraham Flexner suggested in a recent address, the keystone of the educational arch. We agree with him in regarding the improvement of our secondary schools as "the most important work for American education from top to bottom."[152]

[151] Meiklejohn, Alexander, *What Does America Mean?* W. W. Norton and Co., New York, 1935.

[152] See *Teachers College Record,* April, 1935, p. 568.

THE MEANING OF MATHEMATICS

"I COME TO BURY CAESAR, NOT TO PRAISE HIM."

By E. T. BELL
Pasadena, California

A master learns. There is a true story of one of the foremost living mathematicians—call him Y—which illustrates the point of much that is to follow. Professor Y is not only a great research mathematician but also a superb teacher, as shown by the number of first-rate mathematicians he has trained.

Some years ago two young women were working for their doctorates with Professor Y. Their long struggle had reached the last stage: they were to present their joint dissertation before the mathematical seminar. The presentation had been rehearsed fully forty times, and it was agreed that the more confident of the two candidates should present the argument. During the presentation Professor Y sat motionless in the back of the crowded, stuffy room. One theory holds that he slept through the ordeal as the perspiring aspirant covered eight blackboards with masses of horrific equations. Finally the candidate ran out of chalk and sat down. Professor Y sat up with a jerk.

"Have you an eraser?" he asked.

"Oh, yes," the eager candidate admitted. "Is there a mistake somewhere?"

"I don't know. *Rub all that stuff out.*"

Almost in tears both candidates cleaned all eight boards.

"Now, young ladies," Professor Y resumed, "tell us what you think you have done. What does it signify? Are your very beautiful equations only equations, or do they perhaps mean something?"

They did their best, but it was a sorry mess. Ruefully they had to agree with Professor Y that they had wasted two priceless years of their lives in entangling a triviality in a profoundly learned-looking mesh of unnecessary mathematical symbols. They crawled home to a cheerless supper. After the meal they brightened up.

"After all," said one, "he has been most decent to us all these two ghastly years, and he has given us more of his time than we deserved. Can't we do something to show that we are at least not ungrateful, even if we will never be mathematicians?"

"We should," the other agreed. "But what? A gift would look too much like insincerity. I've got it! You know how fond he is of that fat mutt of a dog of his? Let's buy the biggest bologna sausage in town and take it round as a present for the dog. Our dissertation was just about good enough for Y's mutt anyway."

The offering worked wonders. "Ah," the professor exclaimed, "I see there are possibilities I never dreamed of in those young women. I must look into this." He did, and under his expert guidance they finally produced a very creditable dissertation which was nine-tenths his.

Professor Y in his early days had fallen foul of the trap which symbols lay for all mathematicians. Lecturing on a subject which he had not fully mastered, he covered blackboards with equations for twelve weeks, having lost his class hopelessly in the second week. One morning he suddenly stepped back from the blackboard and staring at the cabalistic mysteries in astonishment, he shouted, "Mein Gott! There it is!" The simple demonstration which had eluded him for months had fallen out of itself. That cured Y of relying on symbols. Thereafter he used his head more and his hands less. Let us try to do the same.

What is in it? A fair competence in manipulations is admitted by all to be a necessary prerequisite to understanding a mathematical argument. But no amount of technical facility will of itself teach anyone what mathematics is or what proof means; nor will it suggest what is probably the most important reason why mathematics is today an even more vital human and social necessity than it was in the past. Too much emphasis has been forced upon the practical utility of mathematics "as a tool," to the neglect of the characteristic benefit which an elementary mathematical education can confer on immature minds of normal intelligence. Manipulative skill may suffice for the average technician in the trades or the hack in the lowest ranks of the engineering profession. But it is woefully inadequate as an aid to self-respecting citizenship in even a moderately intelligent society. What shall it profit a mechanic or a surveyor if he can apply the rule of thumb formulas in his handbook as automatically as he breathes if he votes some plausible quack into office merely because he himself, in his practical, efficient blindness, is unable to distinguish between a sound argument and a tissue of rubbish? And what is the social value of a gullible booby who believes every transient fashionable theory in science or economics as if it were the revealed word of God?

These questions are not rhetorical. Unless the student who gets no farther than a first course in algebra or geometry acquires as part of his mentality for life a clear, cold perception of what "proof" means in *any* deductive argument or system of deductive reasoning, his time and effort will have been wasted. Deductive "reasoning" is the subtle device by which the spellbinders dupe their millions. To how many of those who are convinced or converted by some brilliant argument does it occur that the argument, if formally correct, "proves" nothing more than what is implied by the assumptions on which it is based?

The teaching of elementary mathematics, particularly of geometry, can be either a vicious fraud or an unsurpassable object lesson in intellectual honesty. The simple, intuitive ideas of "number" and "space" which most of us have are so "natural," so "necessary," that unless we have once clearly realized that some of them are not natural and possibly none are necessary, we are likely to be easy converts to any quackery based on other "natural" or "necessary" assumptions that may be neither convenient nor sensible. Many of the so-called natural and necessary "concepts" are probably mere conventions which human beings have adopted for their own convenience in understanding the world in which they live.

Even the most elementary mathematics can be so presented that two cardinal facts shall become lifelong acquisitions: (1) *without assumptions there is no proof; (2) no demonstration proves more than is contained in the assumptions.* Elementary mathematics, unencumbered as it is by extraneous scientific or social theories or hypotheses, is the one place in secondary education where these cardinal facts can be acquired. The material to be presented is simple and wholly unobscured by emotions; its lifelong lesson is within the capacity of normal intelligence.

Tradition, or a shovel? It is related of Einstein that he said he was led to the invention of the (special) theory of relativity by "challenging an axiom." The axiom which he challenged was the "self-evident truth" that two events can happen at different places at the same time. Until Einstein upset the axiom by showing that it is neither self-evident nor necessarily true—indeed it is nonsensical—the human race had believed it to be both.

Let us see what the dictionary has to say about axioms. In the unabridged Webster's New International "axiom" is defined in "Logic & Math." as follows:

A self-evident truth, or a proposition whose truth is so evident that no reasoning or demonstration can make it plainer; a proposition which it is necessary to take for granted; as, "The whole is greater than a part"; "A thing cannot, at the same time, be and not be."[1]

This gives a fairly comprehensive summary of the major misconceptions which have been held in the past regarding the mathematical status of axioms. Detailed comment seems unnecessary. The first example, about the whole and part, is particularly unfortunate; for it is neither necessary nor true in a vast region of mathematics.

The definition contradicts the whole history of modern mathematics. It might have satisfied Plato or Euclid, but it should satisfy no one who was born later than 1826. The phrase "which it is necessary to take for granted" is inexcusably misleading. It has not been proved that it is *necessary* to take for granted *any particular* assumption in mathematics. Any assumption that *is* made can be challenged. The whole spirit of the definition is out of date by at least a century, and it is long past time that the word "axiom," with all of its disreputable historical associations, be thrown out of elementary textbooks and replaced by "postulate"—as in modern mathematics.

Before proceeding to a closer consideration of postulates, let us pause here and ask ourselves why mathematics, of all sciences the most progressive and the most prolific in its research activities, is the most backward pedagogically. Who but a demented reactionary would teach physics to boys of fifteen out of Aristotle's "Physics"? Yet the equivalent of that unthinkable stupidity is precisely what we do in geometry. Is it necessary always to plod after poor old dead and gone Euclid, slavishly imitating his primitive attempts to think straight, as if the golden age of the nineteenth century had never been? Shall we never shake off Greek tradition in elementary

[1] The Second Edition of Webster's New International Dictionary, Unabridged, 1934, *almost* sets the matter right, as follows: "A self-consistent statement about undefinable objects which form the basis for discourse. Thus the statement that there is one and only one straight line passing through two given points is an *axiom*." [For 'undefinable' read 'undefined.']

teaching and realize that geometry, wherever it is *used* outside of trivial "applications" for which no formal schooling is necessary, has known the supple freedom of algebra since Descartes, in the seventeenth century, ripped off the strait jacket of Euclidean "demonstration"?

Whoever maintains that we do not continue plodding after Euclid may be asked to produce any textbook on elementary geometry which is fairly widely used in the schools of America and which is less objectionable from a modern standpoint as "a training in reasoning" than Euclid's *Elements*.

This brings us again to the question of what elementary mathematics is all about and what any normal boy or girl can hope to get out of it. Beyond the few trivial applications, what is there in elementary geometry but a training in deductive reasoning? Nothing. And unless the training is modernized, the habits of "reasoning" which are drilled into the pupils are about as bad as they could be. The pity of it is that a decent job would be no harder to do than the awkward muddle consecrated by tradition and sanctioned by mental inertia. Endless generations of committees on the teaching of geometry have proposed timid patches here and there on a corpse that has lain in state for generations. What they need is a shovel and the courage to use it. We can at least bury that rotten word "axiom."

The nature of a mathematical system. Having buried axioms, let us do likewise with another old stand-by which we have inherited from an outmoded way of thinking. There are no "laws" in mathematics. Laws are for lawyers, and popularizers of science who cannot think straight. In mathematics there are undefined elements, postulates, propositions, and theorems.

We start with certain undefined elements. In geometry, for example, two such are "point" and "line." These are probably the simplest two which it is expedient to take in elementary geometry. If we define a "straight line" as "the shortest distance between two points" we are sinning against clarity. What is "distance"? And how are we to know whether one "distance" is "shorter" than another? Or that there *is* a "shortest"? The "shortest distance" definition is at the root of the inability of some of the older physicists to understand the theory of relativity. In their youth they were so thoroughly miseducated in definitions that do not define that they are incapable for life of understanding anything new and sensible.

It is not necessary yet to ask where the undefined elements come from, or why we are moved to select a particular set. Similarly for the postulates. Let us concentrate for the present on the mathematical game itself; after we have seen how it is played there will be time enough to inquire into its origins and our motives in playing it at all.

The undefined elements having been agreed upon, the next step in constructing a mathematical system is the laying down of postulates concerning the undefined elements. Once and for all it should be realized that a postulate is a pure out-and-out assumption. It may have been suggested by "experience"—a more or less philosophical term about which it is difficult to be precise—or it may have been chosen on the mere whim of some mathematician interested in seeing what he could make. In no sense are the postulates "eternal truths" or "necessary"; nor are they guaranteed by any extrahuman necessity or supernatural "existence." The laying down of postulates is a free act of human beings and is not necessitated by any mystical harmony supposed to reside in the eternal essences or superhuman truths of the metaphysicians.

Let us suppose that we have agreed upon some set of postulates for the undefined elements. One postulate for our "points" and "lines" might be "two points determine a line"; another, "two lines determine a point." The latter, by the way, would not usually be admitted in school geometry, for in that subject there are the exceptions introduced by parallels which, by definition, are lines having "no point" in common. But if we introduce an "ideal" point at "infinity"—all a matter of words without any clutter of mysticism—the postulate becomes "intelligible" without any exceptions.

Thus far we have the undefined elements and postulates about them. To the postulates we now apply common logic, or "the laws of thought," and see what the postulates imply. For this we may *assume* that the postulates are "true"—they are true only for the purposes of the deductions we hope to make, and not in any supernatural or mystical sense of eternal verity. The three so-called "laws of thought" of Aristotle are: "*A* is *A*" (the "law of *identity*); "nothing is both *A* and *not-A*" (the "law" of *excluded middle*); "everything is either *A* or *not-A*" (the "law" of *contradiction*). Notice the quotation marks on "law" in each instance. This is to emphasize the fact, which will be mentioned later, that these classical *postulates* of reasoning were once thought to be superhuman necessities and not, as they are regarded today, mere assumptions which human beings have made and agreed to accept. So let us refer to Aristotle's classical "laws" as the postulates of deductive reasoning. Deduction proceeds by an application of these postulates to those of the system—geometry or algebra, say—under investigation.

It is possible to make different kinds of assertions about the undefined elements. The most important of these are the "propositions." A proposition is a statement which is either true or false. A little consideration will show that it is not always easy to recognize whether a particular statement is or is not a proposition. For example, we have an intuitive feeling that "three lines do not always determine precisely one point" is a proposition; it is definitely either true or false, and which it may be is determinable by an application of the postulates of deductive reasoning to the postulates of geometry. But suppose someone were to assert that "truth is more identical than beauty." Is this a proposition? As a matter of fact it is not; it is nonsense. How are we to decide when confronted with a statement in the language of mathematics whether it is or is not a proposition? Since there is no ascertained way of deciding which applies to all such statements, we shall pass on to something easier. But it is worth noticing that even at the beginning serious difficulties arise when we stop to question what it may be that we think we are doing.

Propositions are either true or false. A true proposition is sometimes called a "theorem." If true, we try to "prove" propositions by deductive reasoning. If false, an attempted deductive proof will sometimes reveal the falsity by the "indirect method," which will be noticed in a moment. Proof consists in seeing what the postulates of the system imply. Thus if P, Q are propositions, and if Q follows from P by the postulates of deductive reasoning, and if further it is known (or temporarily assumed) that P is true, then Q is true. In particular, if P is one of our postulates which we have assumed at the beginning to be a true proposition, Q is true. But if it is not known whether Q is true, we may tentatively assume that it is false. If from this assumption we can deduce that Q is also true, we have a conflict with the postulate of excluded middle (in which "A" is now "true"). But we agreed to abide by the postulates of deductive reasoning. To avoid the conflict we say that Q is not false, which we tentatively assumed; namely, Q is true, which we wished to prove or to discover.

The whole game is exceedingly simple. There are but two rules: *state all the postulates; see that no other postulate* (assumption) *slips into a chain of deductive reasoning.* In geometry, for example, it "looks as if" a straight line which cuts one side of a triangle at a point other than a vertex "must" also cut another side. This is the sort of tacit assumption which Euclid or some of his modern imitators might easily make. If it cannot be deduced from the remaining postulates it should be put in plain view with them as another postulate.

Let us dispose here of the tacitly assumed sanctity and "necessity" of Aristotle's "laws of thought"—the three postulates of deductive reasoning. Since 1912 it has been suspected, and since 1930 it has been known, that the second law (excluded middle) is *not* necessary for consistent reasoning.

Now, it may be objected that this pitch of precision is beyond the capacity of adolescents. If that is indeed so, is the alternative to drill them in "reasoning" which is loose, if not entirely nonsensical, from start to finish? As a matter of fact, young people can easily be encouraged to turn their natural tendency to criticize everything and everybody into a most invigorating destructive-constructive criticism of alleged demonstrations in elementary geometry. Encourage them to do everything in their power to catch the author or the teacher slipping into an assumption which has not been stated explicitly as a postulate, and they will soon enter into the fascinating game of close deductive reasoning. Many of the tacit assumptions detected by beginners will be legitimate ones, but they must be disposed of by direct deduction from the postulates or by reference to a previously proved theorem, and not by dogmatic appeal to any book. On the other hand, some of the most vicious assumptions may escape the notice of all except the rarely gifted reasoners.

The game need not begin in all its severity at once—that is a matter for the practical teacher to decide. But unless a student catches a glimpse of the real game he is perhaps blindly attempting to play, before the end of his first year in geometry, it were better for him that he had never begun, and better for his teacher that a millstone had been hung about his neck and he had been cast into the middle of the sea. For what might have been developed into a clear thinker in all probability will have been debased for life into a tippler of patent medicines and a believer in all the quack creeds and popular sciences that keep fakirs fat and prosperous.

From the foregoing sketch of the nature of a mathematical system emerges the distinguishing feature of any such system, which is paradoxically stated in Bertrand Russell's epigram, "Mathematics is the science in which we never know what we are talking about nor whether what we say is true." The postulates from which everything starts are *assumed* to be true; to ask whether they are "really" true is to ask a question which is wholly irrelevant to the mathematics of the situation. The deductions from the postulates have the same "truth value" as the postulates themselves.

Although Russell's remark may tend to overemphasize the view of the older British school that mathematics is identical with logic —a view which, outside of Great Britain, is now generally regarded as untenable—it does call attention to a distinction between mathematics and what is sometimes called "applied mathematics." To see this, consider the statement often seen in elementary texts that the $a, b, c, \ldots, x, y, z,$ of algebra represent "numbers." This statement is sheer nonsense. The letters are mere *undefined marks* or "elements" about which certain postulates are made. The nonsensical statement may, equally nonsensically, be taken as a *definition* of "numbers." To do so is to take unwarranted liberties with language. The very point of elementary algebra is simply that it *is* abstract, that is, devoid of any meaning beyond the formal consequences of the postulates laid down for the marks. Some of elementary algebra is true when interpreted in terms of rational numbers; some of it is false for these same numbers; for example, the statement (which might be taken as a postulate in a first course) that every equation has a root. But we miss the whole point of algebra if we *insist* on any particular interpretation. Algebra stands upon its own feet as a "hypothetico-deductive system." An *interpretation* of the abstract system is an *application*.

Objection may be raised, of course, to any such clear conception of algebra for beginners. But until the student of algebra realizes this conception or its equivalent—as he may easily do after eight months of a first course—he has learned no algebra at all, and to say that he has is simply a gross misstatement of fact. Either we should admit frankly that we are presenting nothing but the mathematically prehistoric and erroneous ideas of the eighteenth century on what algebra is, or we should stop pretending to give the students anything approaching a modern point of view when we are doing nothing of the kind. Would the electricity and magnetism of 1776, for example, be considered as sufficient as the patriotism of that date for boys and girls about to face the world of the twentieth century? If not, why should the mathematics? *Modern ideas are no harder to grasp or to present than are the discarded notions of a century or more ago.*

An example. Geometry is considerably more complicated structurally than algebra. Hilbert's postulates for geometry, one of the best sets extant, number twenty, while some sets of postulates for algebra contain as few as five; one useful set contains twelve. To illustrate what has been said about mathematical systems we shall glance at an elegant set of seven postulates for common algebra, from E. V. Huntington (*Transactions of the American Mathematical Society*, vol. 4, 1903, pp. 31-37). The system defined by these postulates is usually called a *field*, and is identical, abstractly, with common, *rational* algebra. What follows is a paraphrase of parts of Huntington's paper.

The fundamental concept involved is that of a *class* in which two *rules of combination* (or *operations*), denoted by $\oplus$, $\odot$ are defined. *Elements* (members, not further defined) of the class will be denoted by small italic letters $a, b, c, \ldots$. The sense in which $\oplus$, $\odot$ are "defined" is as follows: if a, b are elements of the class, not necessarily distinct, then $a \oplus b$ and $a \odot b$ are uniquely known elements of the class. This is sometimes expressed as "the class is *closed* under the operation $\oplus$, $\odot$." Neither $a \oplus b$ nor $a \odot b$ belongs to the class unless so stated explicitly. These remarks are merely by way of preliminary explanation; the *postulates* follow.

Postulate A1. If a, b and $b \oplus a$ belong to the class, then $a \oplus b = b \oplus a$.

Postulate A2. If $a, b, c, a \oplus b, b \oplus c$ and $a \oplus (b \oplus c)$ belong to the class, then $(a \oplus b) \oplus c = a \oplus (b \oplus c)$.

Postulate A3. For every two elements a and b ($a = b$ or $a \pm b$), there is an element x such that $a \oplus x = b$.

Postulate M1. If a, b and $b \odot a$ belong to the class, then $a \odot b = b \odot a$.

Postulate M2. If $a, b, c, a \odot b, b \odot c$ and $a \odot (b \odot c)$ belong to the class, then $(a \odot b) \odot c = a \odot (b \odot c)$.

Postulate M3. For every two elements a, b ($a = b$ or $a \neq b$), provided $a \oplus a \neq a$ and $b \oplus b \neq b$, there is an element y such that $a \odot y = b$.

Postulate D. If a, b, c, $b \oplus c$, $a \odot b$, $a \odot c$ and $(a \odot b) \oplus (a \odot c)$ belong to the class, then $a \odot (b \oplus c) = (a \odot b) \oplus (a \odot c)$.

The unusual $\oplus$, $\odot$ instead of the familiar $+$, $\times$ are used to prevent any possible misconception that we are talking about numbers as in arithmetic. We are not; the "marks" or undefined elements $a, b, c, \ldots$, are *marks* and nothing more, and the seven postulates state *everything* that we are *assuming* about these marks and $\oplus$, $\odot$. May it be emphasized once more, even at the risk of being tedious, that the word "law" occurs nowhere in the preceding set? There are no *laws* in algebra; there are *postulates* which *we* lay down. Thus, if anyone wishes to name *A1*, he may call it the "commutative postulate" for $\oplus$; he should *not* call it the "commutative *law*" unless he wishes to date himself in the hoop skirt and beaver hat era.

Where did these mysterious postulates come from? Heaven, some will say, and more will think. Deferring any attempt at a more sensible answer, let us stick to the facts and tell the truth: they came out of Professor Huntington's paper. "Yes," some incorrigible mystic may agree, "but where did he get them?" As he does not say, and as it is not ethical in mathematical research to take other men's ideas without acknowledgment, we are driven to the conclusion that he made them up. And that is precisely what he did. There are at least a dozen other sets of postulates, quite different looking from this set of seven, which define exactly the same mathematical system, namely a field. If the reader cares to inspect another set, he will find one by L. E. Dickson containing nine postulates in the same volume as Huntington's set of seven (pp. 13-20).

It is easy to see, as already suggested, that these postulates define common school algebra (including the ban against attempting to divide by zero) up to the point where radicals are introduced. Sets of postulates for radicals (or irrational operations) are also easily manufactured, but there is no need to go into that here. Perhaps the complete freedom, the *arbitrariness* of what we are doing will be more obvious when we realize that *the seven postulates are independent of one another.* That is, it is possible to exhibit a system which does *not* satisfy *any particular one* of the seven postulates, but which does satisfy the *remaining six.* The reader may easily verify that the set of all positive rational numbers with $a \oplus b$, $a \odot b$ *now defined* to mean (or to be) b and ab respectively, that is, $a \oplus b = b$ and $a \odot b = ab$, satisfies all the postulates *except* $A1$. In the same way, a system satisfying all except $M1$ is the system of all integral numbers with $a \oplus b = a + b$ and $a \odot b = b$. A system which satisfies all *except* D is the system of all integral numbers with $a \oplus b = a + b$ and $a \odot b = a + b$. Instances of the remaining systems required to prove the independence are left to the ingenuity of the reader.

There is sometimes a tendency to think of algebra as "infinite"; we can always go on writing down letters or marks and combining them with what we already have. Indeed if we start with *arbitrary* (undefined) $a, b, c, \ldots$, we get $a \oplus a$, $a \oplus a \oplus a$, and so on, no one of which is identical with any other mark. But a *particular* field, that is, a *particular instance* of a system satisfying the postulates of common algebra, does not have to be infinite. *Any* set of n objects can be made a field by proper choice of *particular* definitions for the *general* or *abstract* $\oplus$, $\odot$ occurring in the postulate system. For example, taking $n = 4$, we adapt the following from Huntington's paper:

$\oplus$	z	a	b	c
z	z	a	b	c
a	a	z	c	b
b	b	c	z	a
c	c	b	a	z

$\odot$	z	a	b	c
z	z	z	z	z
a	z	a	b	c
b	z	b	c	a
c	z	c	a	b

Here the first table states that $a \oplus b = c$, $b \oplus c = a$, $c \oplus c = z$; the second that $b \odot b = c$, $c \odot b = a$, etc. It is an interesting exercise to verify the seven postulates for this *finite* field of four elements.

Attentive study of the set of seven postulates and experiment with them will be incomparably more illuminating than pages of explanation of the meaning of algebra, so we shall pass on to something else. A geometrical masterpiece of the postulational method will be found in Hilbert's *Foundations of Geometry* (1899, seventh edition, considerably augmented, 1930).

The Greek tradition. The search for ultimate truths in mathematics is as futile as it is in science. Although mathematical theories or systems do not experience fundamental revolutions with anything approaching the frequency of the like in the physical sciences, nevertheless they do experience such revolutions. Perhaps "revolution" is not the right word in either case. In mathematics broader points of view are constantly being acquired, and what once seemed to be an isolated set of theorems is included as a mere detail in the more inclusive outlook. In physics a hypothesis is patched till there is more patch than hypothesis, when the whole is thrown away, except possibly the dead language in which the outworn hypothesis was described. Even today some physicists find it convenient to speak of strains and stresses in the nonexistent ether which many of them once believed in as anything from an elastic wax to a subtile gas. Mistakes in mathematics have been made and believed in for generations; so gibes at scientists by self-righteous mathematicians come with an ill grace. The fact that mathematics does at least grow continuously and at a terrific rate is sufficient evidence that it evolves and is not the static, lifeless idol of uninteresting perfection that many believe it to be.

Anyone who teaches mathematics, who uses it, or who attempts to advance it either pedagogically or scientifically, will do well to keep in mind that the golden age of mathematics began in the nineteenth century, and not in Greece or in the Europe of the post-Renaissance period. Vastly more was accomplished in that one century than in the whole of preceding history. The first steps were of course slow and possibly more difficult than the achievement of the great century, but it is a mistake to suppose that no pioneering as bold as any of that of the old masters was done from 1800 to 1936. Does it not seem rather a pity and rather inexcusable stupidity that this golden age of the modern maturity of mathematics as a science has left only inappreciable traces on the teaching of elementary mathematics?

The sciences boldly present to beginners in the subject some of what they have discovered that is new, interesting, and useful. Mathematics might easily lead them all. Why, then, does mathematics stumble along in the rear, falling over its ridiculous toga or whatever the absurd garment was that the Greeks favored to prevent them using their legs? Are we everlastingly to have "our debt to Greece" dinned into our ears and do nothing to discharge it once for all and forget it? As long as we continue to believe that the only way of mastering elementary mathematics is to follow the historical tradition we shall exhibit in our teaching and in our thinking the recapitulation theory and fail to get beyond the hairy ape stage of rational evolution.

If these remarks seem too strong, we need but reflect on the months of wasted effort that go into the mastering of numerous trivial and utterly useless theorems in elementary geometry, in the alleged cause of training in deductive reasoning. By the age of sixteen any intelligent boy should be well started toward a *working* knowledge of the calculus. *That* may be of some use to him; fallacious "proofs" that two planes determine a straight line will be of no use whatever to him or to anyone else. And what could be a sillier waste of time than mastering a Greek proof that the circumference of a circle is a constant times the diameter? Why not begin to learn directly how modern human beings do these things? As the author of *Calculus Made Easy* put it in his slogan, "What one fool can do another can." And we might remember that few young people of sixteen are the fools some older people of thirty to sixty imagine them to be. They are young, their minds are fresh, their taste is too keen for stale tradition. They have the capacity to learn provided we have the knowledge and the capacity to teach. The new, the living thing is no more difficult than is the old and dead. It is certainly less repugnant to young minds.

Mathematical invention. Picking up a remark dropped in the last section before our dislike of dead traditions ran away with our discretion, let us try to see how mathematics continually surpasses itself. Incidentally we shall see at least one origin of mathematical systems. This in turn may suggest why mathematics is useful in some of the sciences.

The evolution of the theoretical part of a physical science closely parallels that of a mathematical system. A very brief summary of certain features in the development of theoretical mechanics, electrodynamics, and the quantum theory will suffice to bring out the relevant details.

From the first crude attempts of the pioneers to analyze the motions of material bodies evolved the mystically anthropomorphic notion of "force." Galileo and Newton brought this phase to fair perfection in the statement of the three "laws" of motion. As these laws provide an excellent example of loose definition, let us recall them. First it is necessary to "define" *time* and *matter*. Time, according to Newton, is "that which flows evenly"—a strictly meaningless phrase, because it involves an infinite regress in the notion of "flow" which is itself much later defined in terms of time. Matter is that which "occupies space." Thus space for Newton had an existence independent of any matter which might or might not "occupy" it. Both definitions are "intuitively correct" abstractions of human experience which we *feel* to be sensible and correct idealizations of what we observe. But neither definition defines anything; nor will either bear even a superficial analysis. Both belong to an outmoded way of thinking.

The three "laws" of motion presuppose a clear perception of the *space*, *time*, and *matter* involved in these definitions which do not define. They also presuppose a knowledge of what *motion* is. Motion, in Newtonian mechanics, is "rate of change of position," and is measured, if uniform, by the number of units of "space" passed over in a unit of "time." Is it any wonder that some physicists have an ineradicable dislike for "definitions"?

The three "laws" can now be stated. (1) Every body will continue in its state of rest or of uniform motion in a straight line except in so far as it is compelled to change that state by impressed force. (2) Rate of change of momentum is proportional to the impressed force and takes place in the direction in which the force acts. (3) Action and reaction are equal and opposite.

In (2), momentum is defined as the number of units of "mass" in a body multiplied by the number of units of velocity in its velocity. "Mass" is the "amount of matter" in a body. Although it may seem incredible to some that human beings could ever have imagined that the above statements of these laws mean anything, everyone will recognize that they do give an intelligible picture of human experiences. The enormous amount of dead literature on the interpretation of these laws is one long controversy over the nonexistent "meaning" of the "laws."

In a modern treatment the valuable part of this anthropomorphic legislation of natural phenomena is retained; the laws are accepted as a highly idealized summary of certain simple facts of observation. This establishes the contact between the deductions from the laws and experience. But the statements of the "laws" themselves are shorn of their meaningless mysticism and are put in the form of definite, precise *postulates* to which deductive reasoning can be applied. "Masses" are replaced by numerical constants which are connected by clear-cut postulates with the undefined elements of the system, namely "particles." A particle has both mass and "space-time position." The latter is no mystical concept but is merely a set of four numbers, x, y, z, t arranged in a definite order, say (x, y, z, t), which give the co-ordinates x, y, z of the particle, with reference to some fixed set of axes, at the time t measured on some standard clock. Instead of a circular or meaningless definition of velocity, we now *define* velocity in a given direction, say that of the x-axis, at a given instant t_0 as the derivative of x with respect to t evaluated for the value t_0 of t. This is a perfectly definite mathematical idea. Similarly for all the other mystical notions occurring in the expression of the "laws." Finally, these "laws" are restated in precise mathematical form as *postulates*. The last trace of anthropomorphism and supernaturalism has been sponged from the slate, and we can proceed to apply strict deductive reasoning to this set of three nonmystical *postulates of elementary mechanics*. We have passed from the idealization of experience to a mathematical system. Should we continue, we should presently find it convenient to introduce further *postulates* (*not* "laws," for the hundredth time); for example, the conservation of energy, as an extremely idealized abstraction and extrapolation of observation and laboratory experience.

The history of the classical electromagnetic theory has been similar but more condensed. The theory of the electromagnetic field is summarized in Maxwell's equations. In textbooks which are still widely used, elaborate alleged deductions of these equations from hypotheses which are assumed to be closer to experience than the simple equations themselves are given, with the avowed intention of clarifying their physical meaning. These deductions are cluttered with historical material that is no longer useful or even meaningful, precisely as parts of elementary geometry are clogged and stifled by useless traditions. A modern presentation of the electromagnetic theory acknowledges from the beginning that supposed "deductions" of Maxwell's equations are either circular or fallacious, and boldly states the equations as *postulates* (again!) from which the mathematical theory of the field is to be deduced.

In the modern quantum theory the transition from mystical and unnecessary hypotheses to the mathematical system was even more rapid and direct. The equations of Dirac, say, instead of being "justified" by a nonsensical argument which could justify nothing, are laid down as postulates of the theory at once.

These three examples illustrate the accelerated speed with which the modern point of view has penetrated scientific thought. To modernize our attitude toward mechanics took all of two centuries; electromagnetism about seventy years; the quantum theory about six years. If the penetration of the modern attack has done nothing else, it has at least rid science of the pernicious supernaturalism inherent in the word "law" as applied to science. Reputable science no longer has "laws"; it has instead postulates and hypotheses. The postulates are selected to suit the convenience of the moment, and any or all may be modified or rejected at a moment's notice to accord with growing knowledge of physical facts. Similarly for hypotheses and the mathematical systems—"theories"—constructed upon them. All this emphasizes again the purely human origin of postulates, hypotheses, and mathematical systems. The laying down of postulates in science is a purely human activity, and likewise for the rest.

In one respect a scientific theory differs markedly from a mathematical system. From what precedes, it is seen that the framework at least of a physical theory of the kind discussed is a mathematical system. The abstract mathematics when developed frequently leads to predictions of unexpected physical phenomena. When these are sought and found experimentally, the theory is said to have checked with observation. But if the theory predicts phenomena which conflict with experiment, it must be either modified (in its initial postulates, as a rule), or discarded. Now, although the physical theory has failed, the mathematical system embodying the theory has not; it is in fact precisely what it was before, a set of deductions from postulates expressed in mathematical form whose agreement or disagreement with physical fact is wholly irrelevant to the consistency of the system.

How is it with mathematical postulates and the systems deduced from them? As for the physical theory, the postulates may have been suggested by induction and idealization from observation. This view is held by some to explain Euclid's mysterious definitions of point and straight line. The supposed process is so familiar that it need not be described here. But, precisely as with the "laws" of motion, no progress in deductive reasoning can be made about "points" and "lines" until postulates free of all appeal to physical "experience" are laid down. And who does this? We ourselves.

Let us get it out of our heads once and for all that the postulates of geometry were inspired by some superhuman entity called Eternal Truth existing forever above and beyond our poor human efforts to *create* it. We do as we please about "truth," making our own mathematical postulates and agreeing to use a particular set of rules, called the postulates of logic, to deduce consequences from our freely created postulates. If the scientists of the eighteenth and early nineteenth centuries had but realized that their *postulates* were that and nothing more—not imposed from without by some mysterious lawgiver, they might have written more sensibly than some of them did. The "given laws" of the universe which they imagined they were "discovering" were their own creations in attempts to correlate their experiences in convenient, usable forms of abstract reference. With the conception of science as a social activity of human beings instead of a supernatural necessity to guide them, they would not have been half so disturbed as they were when new experiences toppled one after another of their "eternal verities" and "supreme laws" into the ditch.

Geometry has had the same experience. We shall glance at this immediately. But first let us try to realize that the traditional antihuman theory of the superhuman quality of the "truth" resident in mathematical systems is still widely held by mystics and by some mathematicians who find the history of their subject incomprehensible. What, if anything, is wrong with adhering to this particular superstition? Nothing much, perhaps, except that the irrational believer is likely to come an awful cropper tomorrow or the day after when some researching mathematician shows that a particular pet "necessary truth" is certainly not necessary and is only nonsensically "true."

The classic instance is that of the invention in 1826 of non-Euclidean geometry by Lobatchewsky. There is no need to go into the familiar story in detail. By denying the truth of Euclid's parallel postulate, which had been accepted as an "axiom" of "space" in the most vicious dictionary sense of both much-abused words, Lobatchewsky invented a perfectly consistent geometry which was adequate for any purpose to which Euclid's system had been put. That was but the beginning. Since 1826 innumerable useful geometries have been invented by mathematicians, either to serve definite scientific or mathematical purposes, merely or for pure whim.

The same story was repeated on an even more extensive scale in algebra, analysis, and the theory of numbers. These matters are familiar to anyone who has had a college course in mathematics; so we may pass on to the moral of it all.

It will be noticed that the word "discover" has been studiously avoided in favor of "invent." Did Lobatchewsky *discover* his non-Euclidean geometry, or did he *invent* it? You can get either answer. If "discover" is right or convincing to you, then you believe that there is some sort of fairyland, just as Plato did, where all the theorems of all the mathematics of the year 5000 A.D. are now (and forevermore) hanging like dolls, fire engines, scooters, and replicas of Mickey and Minnie on the unaging Christmas tree of Absolute and Eternal Truth. Step right up, little one, and get the nice shiny theorem which Papa Santa Claus has been reserving for you since ages and ages before the last of the dinosaurs curled up in the mud and smothered to death. Teacher will be delighted to see it, and you may get an *A* on your term examination paper. If you believe any of this, no one on earth can *prove* that you are believing in something that does not exist. All that anyone can do is to point out that you are making use of a *postulate* which will not help your *mathematics* a bit, and possibly suggest that what you really need is a close shave with Occam's razor. But if on the other hand you believe that Lobatchewsky *invented* his geometry, and Euclid *invented* his, you are not likely to believe that Mr. Henry Ford's ingenious mechanics pick their flivvers off Christmas trees. They don't. Still, the mystics maintain that they do. Who is right? Is anybody? Possibly the question is meaningless. But it seems more economical to avoid useless postulates.

Even beginners get a sense of creative power out of simple exercises—such as some of those reproduced in connection with the postulates for a field—when encouraged to manufacture their own mathematical systems. The lesson to be learned from all such exercises is that mathematics is a social activity, a creation of human beings for their own needs from the practical to the aesthetic, and not a dull tyranny imposed upon them. Nothing *forces* us to create mathematical systems; we do it because we like to do it and because we have found it both useful and amusing to mathematicize our outlook on the universe.

Mathematics applied. Many of those who are most vociferous in their mistaken dogma that the only social or educational value of mathematics is its use as a "tool" never even saw mathematics used as a tool for anything more difficult than what a moronic mechanic could do with his thumb. They overlook the fact that there are too many rule-of-thumb mechanics, plasterers, paper-hangers, plumbers, surveyors, and engineers (as well as many first-class men in all of these activities) all unemployed. In their blind enthusiasm for the idiotically impractical they continue to demand that the schools shall produce more of the same. The occupational absorption of the semicompetent probably passed its peak twelve years ago. If the "tool" enthusiasts would catch up with the times and acquaint themselves with some of the enterprises in which a mathematical training is a prerequisite to reasonable technical skill, we might hear less emphasis being hammered on poor old antiques —like the solution of triangles—that went out of date in practical technology a generation or more ago. Instead of wasting weeks acquiring an old-fashioned skill in some trade or technique that was in vogue in the days of the stage coach or early railroading, the student might use his precious time in *beginning* to learn some mathematics that is likely to be of use to him in this decade and the next. The old techniques are all reduced to rules of thumb—as they should be—to make time for something more vital. Let some enthusiast for everlasting drill on solving triangles—useful enough perhaps to our grandfathers—by the textbook methods try to hold down a job in a modern engineering office. The moment he reaches for his logarithmic tables—perhaps even his antiquated slide rule—the scandalized boss will fire him on the spot. Machines have, rightly, replaced brains where brains are unnecessary.

The real uses of mathematics are not in any of this prehistoric stuff. If we must harp on the utility of mathematics in our rapidly changing society, let us try to show that mathematics is useful—indeed indispensable—in vast regions of human activity which are of more vital interest to our race today than all the bookkeeping, surveying, navigating, crude "practical" engineering, and the rest of the paraphernalia that were so dear to the pedagogues of the Muddled Ages preceding the last scientific and industrial revolution —the one we are enjoying now.

The *practical use* of mathematics is in its applications to science. In the physical sciences mathematics is as indispensable as language. In the life sciences, including parts of psychology, it is also indispensable though of less frequent occurrence. It is part of a modern teacher's duty (it seems to me) to make himself acquainted with some of the definite, living uses of the mathematics he professes to impart. A boy or girl growing up today in Victorian ignorance of the part played by science in modern civilization is about as competent to face life in a highly "scientific" civilization as a dirt-eating Indian. And no real comprehension of the rudiments of elementary science is possible without a fair working knowledge of the calculus. If European boys and girls can get at the calculus by the age of sixteen, why cannot American? Surely the answer is not that our own children are stupider than our neighbors'.

Last, of the strictly "practical applications" of very simple mathematics, we may mention what is called "The Mathematics of Investment." Three months of that will not convert any young lamb into a predatory stock gambler, but it will go a long way toward preventing him from surrendering his fleece to the first stock or bond salesman who tells him he has nice soft wool. With a few of the elementary principles of simple and compound interest under his thatch he need fear no Big Bad Wolf.

There is, however, another direction in which mathematics can be applied and this, possibly, is of more importance than the mere bread-and-butter applications. Educators are among those who advocate most loudly a proper training to enable human beings to make good use of the more abundant leisure which, we are told, is to be the common lot when we escape from our present muddle and work fewer hours a day. To enjoy that leisure a modicum of trained reason may not too extravagantly be proposed as an antidote against the boredom of eternal stupid games. The exercise of what intelligence and reasoning power we may have is a more durable form of entertainment than any of the tedious substitutes for thinking which have been invented as time-killers.

Only those who are themselves intolerably dull really believe that the average boy or girl must everlastingly be doing something with a ball or a deck of cards to fight off ennui. A hint of the unknown, of the unaccomplished, will engage any normal group in eager discussion and constructive effort. Only dolts are devoid of intellectual curiosity.

The natural thirst for natural knowledge can be encouraged into a lifelong taste. This thirst is waiting to be capitalized into a social asset by a more vital training in elementary mathematics than that which is now offered to boys and girls of twelve to seventeen or eighteen. Without a working knowledge of the elements of the calculus the seals of the great books of physical science must forever be unbroken to those who would try to catch a glimpse of what modern thought imagines the universe to be. A good start is all the formal instruction necessary. That and an awakened curiosity should suffice to master the competent expositions that a more interested generation will undoubtedly demand. But even if this much should be too utopian for the immediate present, there is no reason whatever for allowing generation after generation of school children to finish their education ignorant of what mathematics is and almost wholly uncritical of arguments presented in the form of deductive reasoning.

To sum it all up: the future function of an elementary mathematical education should be to fit minds to the scientific twentieth century.

Modest proposals. Although it may be outside the topic which was assigned to me, I should like to make four concrete proposals in the form of suggestions for a definite program to revise elementary instruction in mathematics. Dean Swift, it will be remembered, once made a "modest proposal" to relieve the Irish famine. Swift failed to relieve anything but his own "savage indignation" because his thick-headed superiors thought he was only joking. They laughed at his macabre joke and let the Irish keep on starving. To prevent a similar misunderstanding, I emphasize that although my own modest proposals may seem fantastically utopian in the present state of mathematical education in America, I mean every word of them. Further, I see no great merit in always patting ourselves on the back for the excellent job we are already doing in secondary mathematical education when a little extra effort and a little more ambition would make it a very much better job. Here are the proposals.

1. *Make it ruinously unprofitable for authors to write antiquated textbooks and for publishers to produce them.*

To accomplish this I suggest the next two measures, the first of which is destructive, the second, constructive.

2. *See that the professional journals of secondary education in mathematics obtain and print thoroughly competent and fearlessly*

critical reviews of all mathematical texts at present used in the schools and of all new texts as they appear.

For example, if a new geometry appears, let the first twenty propositions (usually the first four will suffice) be analyzed. If a single unstated assumption is used in any proof, the book should be forthwith banned or revised. A modern treatment or none, of both algebra and geometry, is what the student of today needs.

3. *Let a large number of both the younger and the more experienced teachers of elementary mathematics take it upon themselves to produce textbooks which will meet the demands of strict deductive reasoning* (in addition, of course, to the necessary formal technique).

This might be accomplished by more teachers availing themselves during their summer vacations of the opportunities provided *in the departments of mathematics* of the larger universities for becoming thoroughly familiar with *modern* presentations of *elementary* mathematics. I have no doubt that most of the progressive departments of mathematics in the country would be glad to co-operate.

4. *Aim at a working knowledge of the calculus as the crown of a secondary school education in mathematics.*

THE CONTRIBUTION OF MATHEMATICS TO CIVILIZATION

BY DAVID EUGENE SMITH, PH.D., D.SC., LL.D.
Teachers College, Columbia University

The landscape. If we seek a point from which to view the broad landscape which we term civilization we may proceed by divers paths. The one most readily found and with the easiest gradient leads us to the spot from which we see first of all that region which concerns the affairs of daily life—matters of trade, of finance, of profit, of loss, and all that concerns this machine age which is overwhelming us with social problems which we strive, often with discouraging results, to solve. Whether or not this region is part of the landscape we wish to see depends upon our definition of civilization. To it, however, Mathematics makes contributions so evident that we need hardly consider them. If we imagine, for a moment, what would happen tomorrow if every trace of mathematics were banished tonight from the region of commerce, of transportation, and of daily life in general, the picture would be that of a desert of starving humanity: no medium of exchange, no buying the needs of life, no measuring of time or of objects, no machinery for light or heat, no transportation, not even the simplest barter that required counting. The picture is too impossible for us to conceive, save in some such slight degree.

A second way of approach may be the one leading to a point from which we view the domain of the sciences. This is the more interesting, for the picture is more rapidly changing, science being forever new. With the included region of applications the changes come not merely year by year, but day by day and hour by hour. Compared with the progress of science, the affairs of daily life seem stagnant except as they lay under contribution its applications. Such applications are seen in our means of communication —by travel, by the printing press, by electricity, and by radio waves—and are also seen in the conveniences of the modern home.

It is an easy matter to show how far-reaching is the influence of mathematics upon such subjects as physics, astronomy, map drawing, insurance, biology, and most of the other branches of human knowledge. A glance at a standard treatise upon any one of them above the schoolbook class will show how indebted each is to mathematics, advanced as well as elementary. The landscape seen from this point of view is one of grandeur; and to it we shall return.

A third route leads to the point of view from which to see contributions of Mathematics to the forming of the minds of men—the acquisition of habits of thinking such as geometry may cultivate, and the consideration of what Professor C. J. Keyser has called "the value of rigorous thinking." From this point we see first of all the region of psychology in which we consider the cultivation of habits which transfer to related fields of mental activity. The view is enticing, the more so because the path is over a rocky road and a long one. It is one to be taken by that rather unusual being who makes of philosophy something besides mere words—the man who has the genius of originality and the power of lucid expression.

There is also a fourth route to the point we seek. As we make ascent we may consider the meanings of the words "Mathematics" and "Civilization." Failing to define them satisfactorily, we may attempt to explain them sufficiently for our purposes. We may even branch off to the points of view to which the other routes have led, finally reaching the one where we can see the landscape at its best and with the widest sweep of vision.

The meaning of mathematics. In the first place we should naturally wish to define our leading terms, Mathematics and Civilization, for upon these definitions depends the spirit in which we view the landscape. We shall not succeed in our attempt, however, for the first of the terms changes from time to time, and the second changes from place to place as well.

To illustrate one of the difficulties of definition—until about the year 1500 music had been, for two thousand years at least, classed as one of the branches of mathematics, and the influence of music upon civilization has been very great. With the coming of the Era of Leisure, if such an era really develops, it will be even greater than ever before. When we consider the interest shown in the mathematics of music[1] by such scholars as Pythagoras,

[1] See Archibald, R. C., "Mathematicians and Music," *American Mathematical*

Philolaus, Aristoxeneus, and Ptolemy among the Greeks; Boetius[2] at the close of the ancient Roman period; mathematicians of the Renaissance period and later, like Mersenne, Grammateus, Stevin, Huygens, Jacques le Fèvre (Faber Stapulensis), Euler, and Johann Heinrich Lambert, not to speak of the interest in mathematics shown by musicians, we can see the probable development, in the future, of a still closer union of the two branches of knowledge. The contribution of mathematics to civilization will then have a new, or at least a more modern, meaning.

To say that mathematics is far-reaching in this special field of the fine arts does not, however, help us in our definition of the subject; nor shall we be assisted by the fact that what we in the English-speaking world call "Arithmetic" was practically never considered a part of mathematics until about the beginning of the seventeenth century. Before that time such branches as "profit and loss," partnership, and the computation of interest were not considered as divisions of the subject; nor were the "four fundamental operations" generally so classed. Arithmetic was then the theory of numbers—primes, roots, series, and the various properties of numbers. The name "Arithmetic" began to be applied to computation simply to add a supposed dignity to what had been the training of a merchant's apprentice. We have a similar and rather ridiculous movement at the present time in the effort to apply "Research" to all kinds of merely mechanical study, even down in the primary school. In the seventeenth century the effort to adopt the scientific term by the elementary school succeeded, the old "Arithmetic" being then changed to the "Theory of Numbers." Probably "Research" will have a similar fate, some other term being invented to take its proper place.

Apparently, therefore, we shall have to admit that "Mathematics" is a kind of indefinable term, like "space," "angle," "plane," and "time." We may explain the term with a certain degree of success, but we have no good definition for it. We might say that it is a branch of logic based upon certain assumptions which may or may not be true and which relate to certain imagined things which we call points, lines, planes, solids, and angles, which we also cannot satisfactorily define. What we then wish to know is

Monthly, XXXI, 1-25; Barbour, J. M., "The Persistence of the Pythagorean Tuning System," *Scripta Mathematica*, I, 286-304.

[2] Or Boethius, *De institutione musica*.

its influence upon "Civilization"—a term which is equally impossible of satisfactory definition. For the mythical "man in the street" it may be sufficient to say that Mathematics is the thing which contains such other things as arithmetic, algebra, geometry, trigonometry, and their offspring in college and university courses.

The meaning of civilization. We see that the meaning of "Mathematics" is not easily stated, and it is evident that the word "Civilization" is still more difficult. I have many times talked with scholars of undoubted ability who do not think that the Western World represents as high a degree of civilization as parts of the Orient. They point to our crime, our merciless wars, our literature, and our eternal seeking for wealth, comparing our lives with those they live—their absence of nerve-racking business customs, of salacious fiction, and of greed for money. The contrast is not always in our favor. If I should ask them about the influence of Mathematics upon Civilization, the answer would be quite different from any that we should be likely to get if we asked a Wall Street broker, an automobile manufacturer, or a professor of education in some of our colleges. Their psychology is possibly not ours; nor are their ideals, their scholastic attainments, their use of leisure, or their appreciation of the beautiful in architecture, in painting, in music, or in any of the other fine arts. No one can live among those who look down upon our civilization without realizing the soundness of some of their views. They agree that we excel them in certain comforts of living, in our science of sanitation, in our efforts to better the conditions of the poor, and in such things as scientific agriculture, but they deny our superiority in the finer use of the mind or in general happiness.

They would say that the influence of mathematics upon the Western World relates solely to science, finance, the applications of physics, astronomy, and the like. This is not the case, but it is what appears to many thoughtful men to be true. As to the Orient, the influence relates to abstract values—the cultivation of the habit of logical thinking and the explanation of the universe in which we find ourselves—and relates less to the applications of our scientific knowledge to industry.

Having now considered in a cursory fashion the complexity of the problem, let us turn to some of the more obvious lines of influence of an indefinable subject upon an indefinable condition of part of the human race.

The genesis. The contribution of Mathematics to Civilization began millions of years before Civilization was born. As soon as the earth began to cool from its gaseous state and crystals began to form, mathematics began its contributions. Moved by some force similar to the habits of men, similar to their inherited instincts, similar to the fashions of the human race, these crystals assumed certain forms and seemingly passed them on to their descendants, a kind of perpetual inheritance. Let us break, for a moment, with the conventional treatment of our subject and look at it from a point of view different from the one with which every teacher is familiar.

Still speaking figuratively, Quartz made up its mind to crystallize in the form of a regular hexagonal prism, capped by a pyramid. Calcite or Iceland spar decided on a rhomboidal solid with its interfacial angle about 75°; Alum, Gold, and Magnetite selected the octahedron; Salt and Fluorspar preferred the cube, and in similar fashion other substances took on, in crystallizing, these or other definite geometric forms. Such substances lived their lives, again speaking figuratively, and passed on their habits to their descendants; and many millions of years later humanity came and followed the fashions which some unknown Power—call it "Nature" if you wish—set in an age so far removed from the present that our minds are helpless when they attempt to grasp the length of time which has since elapsed. Throughout these long reaches of time all these crystals have obeyed the law that $F + V = E + 2$, where the letters stand respectively for the number of faces, vertices, and edges. The law may have been known to Archimedes more than two thousand years ago; it was announced only three centuries ago and was made generally understood by Euler about the middle of the eighteenth century; but it has existed for all time because it is one of the eternal truths of geometry in the space in which we speak of living. The fashions of these crystals, and these eternal laws, fixed long before the world's genesis, are followed today in the arts of all ranks of civilization. The world seeks, as it has always sought, geometric forms for its dress, its habitations, and its decorations of rooms, of gardens, and of its festive boards.

When Job (xxxviii, 22) asked, "Hast thou entered into the treasures of the snow? or hast thou seen the treasures of the hail?" he must have felt the beauties of the crystals of the former, and of those which the latter concealed. Ever since civilization dawned, those who saw the snow fall have felt with wonder the beauty of its flakes, each built upon the same general plan but each varying slightly in the minor decorations. We have no means of knowing when the first snow fell upon the earth, or who it was, millions of years later, who first noticed its symmetry or the fact that its foundation is an equilateral triangle, or one such triangle superposed on another to form a regular hexagon. It was not until 1830 that Hessel proved that there are thirty-two types of symmetry possible in crystals, perhaps 50,000,000 years after primitive man came upon Earth. Since 1885 the snow crystals have been systematically photographed, and out of 4,800 photomicrographs no two have been found with what may be called precisely the same frills. There is a rivalry in dress in the snowflake as well as in the dresses of our day. This much is known—that the decorations of the snow vary with the height of the clouds and the intensity of the cold.

To the artists of our civilization all this has an interest, and likewise to our natural scientists. To our mathematicians it has the added interest that the lines from the center to each vertex of the triangle are the graphs of the three cube roots of unity; that is, $1, -\frac{1}{2}+\frac{1}{2}\sqrt{-3}$, and $-\frac{1}{2}-\frac{1}{2}\sqrt{-3}$. If the hexagonal form is taken, the graphs are the six 6th roots of unity.

The child playing with the kaleidoscope unconsciously studies the snow crystal as he watches the beautiful forms—for the name of his toy is made up of three Greek words meaning "beautiful," "form," and "look." As he turns the cylinder he sees what Nature does, and he can make an almost endless series of pictures, all based upon the equilateral triangle, but all differing in slight details. To his civilization geometric forms have made a contribution.

The Race, too, has its childhood in its trivial superstitions, and its imprisonment behind the walls of tradition. Perhaps Nature has the same. Quartz seems to start out with a desire to make a perfect crystal; it ends with the recognition of a slight imperfection. It may cut off some of the angles, or its axis of symmetry may not be exact. There is nothing perfect in the Universe. When Mohammedans build a mosque they purposely leave some slight imperfection. "God alone," they say, "is perfect." They feel that it would be sacrilege for man to attempt to make anything without a blemish. Even their Korans, the most beautifully and carefully written books in the world, seek to have a perfect text, but in the letters or the geometric decorations there will necessarily be some imperfections which may be thought to show their reverence for the Perfect One, their Allah.

It is not without value to speculate upon the reasons why the crystals, in the eons of the early life of Earth, figuratively speaking, decided upon building their edifices in such fashion as to give different rates to heat and electricity as the waves pass through them, the rates varying according to the different directions which they take. It is also interesting to consider why and how the optical properties of crystals vary according to the substance. We may say that all these idiosyncrasies are due to the way in which the electrons, protons, neutrons, atoms, and electric waves arrange themselves, but the original question simply gives place to another: Why did these electrons and their relatives decide upon these arrangements? This we cannot answer as we should wish. We can simply say that Physics finds what is to be measured, for example with respect to crystals, and Mathematics devises formulas which state the law which is discovered. The two can combine to give valuable results, and Commerce benefits by using these in the verification of precious stones and the determination of mineral species. Did Mathematics contribute this incident in the structure of Civilization, or was it Nature? If the latter, what is Nature? Is she the twin sister of Mathematics? The two seem closely linked. Cut through the seed bulb of the rose, or take a cross section through the core of an apple, or study the web for which the spider spins the threads, or under a powerful microscope view the geometric shapes of various bacilli. Which contributes the more to Civilization—Geometry or Nature? At any rate, Geometry does not contribute disease germs as Nature often does.

Contributions in the early historic period. Leaving these speculations about what we call the prehistoric period, concerning the civilization of which our ideas are too cloudy to be called scientific, let us consider the earliest written records which have as yet been revealed. There are several regions which, if we were to proceed with a view to scientific accuracy, should be studied with special care. These are China, the earliest records of which are confessedly of doubtful authenticity; India, of which the early chronology is quite unknown; the lands overrun by the Sumerians, the recent discoveries in which are opening a new vista, putting our knowledge of algebra, for example, back more than a thousand (perhaps two thousand) years before Euclid's geometric treatment of the subject; and Egypt, in which new phases of the early history are even now being revealed.

For our present purposes, however, it is immaterial whether our civilization had its beginning in Central America, in India, in China, in Iraq, or in Egypt. What is concerned with our present interests is the influence which mathematics had upon the early stages of any of these civilizations. When we attempt to weigh this influence we are confronted by the fact that the earliest written records in every case show that the chief interests, aside from mere barter, of the countries mentioned were based upon a study of time and the heavens. This study involved extensive number systems, and the angle measures required in their astronomical records.

Civilization has always sought for the meaning of the Universe, and to this search Mathematics has contributed throughout the historic period at least. The primitive savage went to the "medicine man," the fakir, the fortune teller, the priest, the self-styled prophet, the astrologer, the numerologist, and later, as science slowly replaced superstition, to the astronomer, the physicist, the physician, the chemist, and the natural scientists in general. In the scientific period, each of these authorities has come to the mathematician for assistance, and it is from him that there has come a new international written language of science—the algebraic formula. In physics for example, we have come to depend upon formulas and their graphs as our most potent aids in thinking; and through their manipulation by algebra, the discovery of new laws.

Instead of considering the early theories of the universe in detail, we may turn for a moment to the fascinating subject of the Maya civilization on our own continent. A German astronomer, Professor Henseling, has asserted that a study of the inscriptions, especially at Copan and Naranjo, shows that more than 5,000 years ago the knowledge of astronomy possessed by the Mayan scholars was much superior to that of any other people in the world. Indeed, he contends that the inscriptions reveal a system of chronology which goes back to 8498 B.C. This assertion is based upon data determined by Professor Ludendorff, director of the Astrophysical Institute at Potsdam, who has himself, it is asserted in the press, come to recognize the validity of Henseling's conclusions. Whether or not these dates are accepted, there is no doubt of the early interest of the Mayan people in the mathematics of calendar making and of the approximate accuracy of their work.

The earliest evidence that we have of the contribution of Mathematics to Civilization, beyond such matters as the mere counting needed in barter, and taxes payable in kind, is seen not only in the calendar of the Mayas or the records of celestial movements. It is everywhere, in every continent, connected with religion, being little more than elementary astrology. For example, in China such records as may fairly be called authentic relate to this subject, and the philosophic-religious cult of Taoism originally concerned the revolution of the heavens. The Sumerian records, recently made known, tell the same story; the astronomy of these people was passed on to the Babylonians, and by them to the Chaldeans. It was probably from the latter that Thales learned how to forecast the eclipse which contributed to his fame. So it was that, for many centuries, such ancient races as the Assyrian, Aramaic, and Egyptian brought their mathematics to bear upon that phase of their civilization which is concerned with first causes, with our status in space, and with the movements of the stars which were thought to control our several destinies.

It was not alone in the making of calendars or in the measurements of angles, like those which locate the prominent stars, that mathematics contributed to the early civilizations. The sundial was scientifically made with a fair degree of accuracy in the second millennium before our era, and probably earlier—an important step in the measurement of time and in the consequent civilization of the race. Aristarchus, in the third century B.C. measured the distance to the moon, failing to approximate it closely simply because accurate instruments were not available; by the time of Ptolemy the astronomer, mathematics had opened up a vast field for the speculation of men. How this field has been cultivated by such mathematicians as Copernicus, Kepler, Newton, Euler, and Laplace need not be considered. Eliminate the mathematical achievements of men of this type and the world would return to the mentality of those who placed faith in the astrologers and soothsayers of ancient times.

So it came about that Mathematics contributed to that phase of Civilization which leads mankind to search out the nature of the universe as a whole and which led the English poet laureate thus to express, two centuries ago, the ancient feeling of mankind:

> O God, we thank Thee for this great universe, our home; for its vastness and its riches and for the manifoldness of its life!

There are those who would say that Religion is not a part of Civilization, although their number seems silently decreasing. There is no doubt, however, that Mathematics has contributed to practically all the religions of the world. The earliest authentic records of India show how advanced was the geometry which the priests applied to the building of altars. The *Sulbas* of the Vedic priests of India were concerned chiefly with this work.[3] The altars were of various shapes, such as those with square, circular, and semicircular bases, it being required that areas be the same in all cases. This led to the approximate squaring of the circle and its related problem, the circling of the square. These problems refer to the construction of sacrificial altars and may date back as far as the *Rig-veda Saṁhita*, some 3000 years B.C. This raises the question of the influence of the geometry of India upon the work of the Greeks, as in the altar at Delos and the duplication of the cube.

The vision thus opened suggests other possibilities, chief among them being the mathematics of temples, tombs, and other religious edifices. The contributions of Mathematics to the art and rigidity of the buildings are too extensive to be mentioned in this article. Such a study would lead to the question of the orientation of temples, of the real or imagined mysticism of the mathematics of the Great Pyramid of Gizeh, of the use of the Golden Section in the dimensions of a temple, of the designs of a Gothic church,[4] all of which are contributions of Geometry to some of our noblest evidences of Civilization and its fine arts. The word *Śulba* or *Śulva* is a geometric one and refers to measures, including a rope for measuring lengths. It is also used in connection with a compound word meaning "the science of geometry," and in the texts mention is made of "the rope holder" ("the royal land surveyor"), and the "uniform rope-stretcher." This seems to show that the "rope-stretchers" (*harpedonaptae*) of the Egyptians concerning whom Democritus wrote in the fifth century B.C. were related to the earlier altar builders of India.

To Religion, then, Geometry made a definite contribution extending over many centuries. Of the influence of Astronomy upon the early religious life of all peoples it is unnecessary to speak, since the evidences are well known, as in Biblical passages like these:

[3] *The Science of the Śulba,* B. Datta, Calcutta, 1932.

[4] See the writer's *Mathematica Gothica,* Paris, 1926, privately printed, now in process of revision and elaboration.

The heavens declare the glory of God, and the firmament showeth his handiwork.

He telleth the number of the stars; he calleth them all by their names.

The heavens are the works of thine hands.

When I consider the heavens, the work of thy fingers; the moon and the stars which thou hast ordained; what is man that thou art mindful of him? and the son of man that thou visitest him?

To the nomad shepherds watching their flocks on the slopes of the mountains of Palestine, Iraq, or Iran (Persia) tonight, as they have watched for thousands of years, such thoughts as these must vaguely come, although in simpler language.

All this has little direct bearing upon the contribution of Mathematics to Civilization; but it reveals the early steps in the understanding of the noblest branch of applied mathematics—our present astronomy.

Contributions to present civilization. Referring to the contribution of Mathematics to the Civilization of today, and taking Civilization to include our industries, our financial interests, our sciences, our sports, and our arts, we come to a field of immeasurable extent. The number of applications of Mathematics to the interests of mankind, existent or potential, is too great for our ready comprehension. We may, however, refer to a few by way of illustration.

It is manifestly impossible in an article of this kind to summarize the contributions of Mathematics to that element of Civilization which we call Science. In the first place we are confronted by another term which admits of no definition that is generally satisfactory to the scientists themselves. It will be better to select a single case of the simplest group of sciences and consider a few of the contributions which Mathematics has made and is making in this field. Let us take for our illustration the field of Geography, one which is more extensive than we may at first imagine, and one which is continually expanding, with the aid of Mathematics, our ancient and ever-increasing inheritance. It must be understood, however, that this is one of the simplest illustrations that we could take. If the contribution of Mathematics is great in a case like this, in other sciences it would naturally be expected to be much greater. We shall find, however, that Geography is so extensive that it makes use of fully as wide a range of mathematical subjects as some of the other sciences; in fact the barriers between Geography and the other sciences have been almost eliminated—as in the case also with the other departments which were founded separately by ancient scholars.

Contributions to geography. Probably the word "Geography" will call to the reader's mind a simple subject of study in the elementary schools. It includes, however, what we call geodesy, which has to do with both the size and the figure of the earth—each of which is a mathematical subject. If it is worth while to know, with a close approximation to accuracy, the distance from New York to Naples, to Greenwich, or to the North Pole, we need something besides a measuring tape. We need to know not merely that the earth is not flat, but that it is not even precisely spherical. We must know its radius at different points of our terrestrial spheroid, and the mathematics involved is by no means simple. One who would wish to master the subject must be well versed in spherical trigonometry, the calculus, least squares, and numerical equations of considerable difficulty, besides being an expert in the mechanics of measurement.

Cartography, less formally known as map making, has manifestly much to do with civilization. It may seem to a casual observer to be a simple matter. If, however, the reader will look at even a popular article on the subject, in any standard encyclopedia, he may find that fields of Mathematics which he thought to be mere useless abstractions when he was in college are necessary to any intelligent examination of the subject. In particular, he will find a definite use for curves of which he may have only a faint memory —orthodromic lines, loxodromic lines; for projections of various types—cylindric, orthographic, stereographic, perspective, gnomonic, central, conic, polyconic, zenithal, elliptic, and others. To make use of these he will need a good command of geometry, algebra, trigonometry, elementary calculus, and differential equations. He will, however, find the need for more mathematics than this if he wishes to pursue the subject into the fields of the influence of the tides upon the land surface, the measurement of terrestrial magnetism and its effect upon the instruments employed. He must also be conversant with contour maps, not merely of dry land but of the wet land which forms the bed of the oceans. To master this part of geography will require a mathematical knowledge of modern "soundings" by echoes from the sea bed, leading him into the domain of physics. The seeming miracle of telling, by a seismograph, the time, place, and force of an earthquake, possibly thousands of miles away, is but one of the relatively simple results of a combination of mathematics, geography, physics, and mechanics.

Contributions to sciences in general. What has been said of the uses of Mathematics in Geography, and of the wide field opened by this elementary study of the earth's surface, is only a small intimation of the uses of mathematics in connection with what we call our present Civilization. Science has been rather poetically defined as that branch of knowledge that exists by the aid of Mathematics. It calls upon the latter to determine the number of thermal units per cubic meter of various gases, the yield of gas in cubic meters per metric ton of coal at various temperatures, the height of the stratosphere, the power of the cosmic rays, the size and frequency of sun spots, the equations of chemical compounds, the formulas for theoretical gravity at the surface of the earth, the formulas needed in the Mendelian theory of heredity, and those in hundreds of other branches which compose the tree of science in the present time.

From the formulas created it leads us to their pictures—the graphs—and there are few statistical tables which do not now make use of these visual aids. The man of commerce, the financier, the engineer, the statistician—all these and men of many other gilds rely upon the graph not merely as an expression of results but, figuratively speaking, as an important element in the thought process.

It is not necessary to attempt to recite all the contributions of Science to Civilization. The two terms may almost be called synonymous, since Science enters into all the arts and even into the modern concepts of Religion and Philosophy.

It should be understood, however, that there is a line of separation between Mathematics and Science, and hence between Mathematics and Civilization. As already remarked, Mathematics says to Physics, Medicine, Chemistry, and Astronomy, "If your observations are correct, then this formula is correct, and hence this numerical result is correct." In other words, Mathematics is always based upon assumptions. The old cosmologies made such assertions as this: "There are seven hemispheres in the heavens; they move in a certain way; now tell us when planets *A*, *B*, and *C* will be in conjunction." Mathematics then takes up its rôle; it computes results from the data provided. If these data are right, the results will be right. Although the old cosmology was wrong as a whole, it was approximately correct in some of its observations, and hence the mathematical results were equally approximate. Professor A. N. Whitehead, in his *Adventures of Ideas* (1933), makes this interesting assertion: "The Certainties of Science are a delusion. They are hedged around with unexplored limitations." And so were the ancient cosmologies. The mathematics was not at fault; it was the observations and the confused thoughts of the astronomer that were the source of inaccuracies of result. Newton based his theory of gravitation on the observations made by himself and men like Cassini; his mathematical law was only as perfect as these observations were perfect. Einstein modifies it, and his modification is perfect only so far as observations conform to or serve as bases for his new formulas.

In Science we are especially concerned with the probability of our conclusions, and hence the study of this branch of applied mathematics is a matter of importance. For this reason some reference to it is desirable at this point.

Probability as a factor in civilization. Mathematics is constantly influencing Civilization through the Theory of Probability. It lies at the basis of all types of insurance and also of statistics when used in prognostications, such as the probability of a rise in stocks, of good or bad crops next year, and of an eclipse within, say, 0.1 sec. And yet its laws are based upon observations which are never absolutely exact. We assume, however, that the observations are exact and frame mathematical laws to represent the assumptions. From this time on, the solutions of equations based upon a given law are correct; but although the mathematics is exact the results are no more correct than the physical assumption.

It would almost be justifiable to say that the probability that a probable statement is probable, is very improbable—paradoxical as this may sound—although even in finite time and space the mathematical treatment of it may be exact. As Bertrand Russell has remarked, "the concept of probability is wrapped in obscurity and affords the chief scandal of modern logic." In spite of all this, however, the mathematical treatment of a formula—which in itself is logical or illogical, true or false—stands for accuracy. It should be repeated, however, that we have to bear in mind that Mathematics is based upon certain assumptions which have been called by various names such as *petitions*, *assumptions*, or *postulates*, and the mathematician, considering a concrete problem, pays little attention to whether or not they are true in infinite as well as finite space or time. All that Mathematics says is this: If this postulate is true, then my results are true. Indeed, a familiar definition of mathematics is expressed in these words: "If *A*, then *B*." Postulates are based upon observation; mathematics is based upon reason. To Mathematics it is of no matter whether Newton's Law of Universal Gravitation is true or not. It asserts that if two particles have masses M and M', the force F of their mutual attraction, when they are d distance apart, is given by the formula $F = GMM'/d^2$, where G is a universal constant known as "the constant of gravitation." Given this formula, which expresses in algebraic shorthand the law, Mathematics operates with perfect certainty to find the value of any letter in terms of the others. Similarly, Mathematics does not find the precise distance from the earth's center to that of the sun. It simply says that, assuming that the observations, on for example the transit of Venus, are correct to a certain degree of accuracy, then the distance, to a corresponding degree of accuracy, is exact.

The temptation of teachers is often to question the value of such topics as determinants, differential equations, convergency tests, recurring series, summation of series, series in general, various formulas for means, the theory of functions in general, elliptic functions, various periodic functions in particular, inequalities, moduli, differential equations, binomial equations, and limits. Of course most of these topics are beyond the high school range, some are beyond the elementary college work, but this has no bearing upon our problem—the influence of Mathematics upon Civilization in general. It would not be contended that the insurance of life or property and insurance against accidents is not an important factor in our modern civilization. Indeed, without this factor our whole problem of security of home and family would remain unsolved. It will probably be agreed, therefore, that the mathematics of insurance has a decidedly important bearing upon our Civilization. It may be said that the amount of mathematics required for this purpose is slight, but not long ago I listened to a lecture by Professor Fréchet of the Faculté des Sciences of the University of Paris. It was upon a certain type of probabilities as developed by such leaders as Frobenius, Markoff, Poincaré, Hadamard, Hostimsky, and himself. In this lecture all of the above-named topics were freely used in the development and use of the necessary formulas of the science. Among the scholars who made up the audience were representatives of insurance companies of various kinds, statisticians, and financiers, and in the discussion the speakers representing these interests stated that the results set forth in the lecture were freely used by them, even in what would seem to be the very elementary field of preparing tables for use in the subject of fire insurance.[5]

Conventional contributions. The references to the contributions of Mathematics to Civilization have, in this paper, been thus far somewhat unconventional; that is, many of them have referred to matters not ordinarily considered as the application of mathematics to everyday life. The branches of Mathematics needed in one of the sciences usually thought of as merely a school subject have been mentioned, as also those needed in such a more advanced subject as the Doctrine of Probability. These two regions have been selected as types, and attention has called only to the most apparent applications of mathematics to these subjects.

It should be observed, however, that a consideration of such subjects as Accounting, Aeronautics, Astronomy, Astrophysics, Atomic Theory, Ballistics, Bridges, Calculating Machines, Chemistry, Engineering, Geophysics, Kinematics, Light, Measures, Optics, Perspective, Physics, Sound, and the like, will give some idea of the extent of the contributions which Mathematics is making to the arts and sciences. The reader who cares to get a synoptic view of the subject would do well to consult the *Encyclopaedia Britannica* (14th Edition) under the above heads, following this by a further examination of articles of a similar nature. After such an examination it will assist in the comprehension of the importance of mathematics in our time if the reader will consider what would happen to modern civilization if all mathematics, and the results which it has given to the world, were absolutely destroyed.

It is also the conventional practice to call attention to the older contributions of Mathematics to the Civilization of today. Although these may be found in any standard history of mathematics, it may be of assistance to the reader if a few of them are briefly mentioned.

In the fifth century B.C. three Greek philosophers—Zeno of Elea, Leucippus, and Democritus—were connected with the introduction

[5] See also Struik, D. J., "On the Foundations of the Theory of Probabilities," *Philosophy of Science*, I, 50-70.

of the idea of infinitesimals. It seemed a useless subject, but it was upon this idea that Newton, two thousand years later, based one of his theories of the calculus. In the same century Antiphon, and in the following one Eudoxus, improved upon this by introducing the method of exhaustion, which still more nearly approached the modern calculus. In the third century B.C., Archimedes approached it even more closely and practically made use of integration. No one of these men conceived of the tremendous importance of the calculus in the present civilization, as in physics, mechanics, astronomy, and other sciences, but each laid out the first stages of the route which Newton, Leibniz, and their successors followed, and which eventually assisted powerfully in placing science where it is today.

The Greeks knew a number of important curves, such as the quadratrix, the conchoid, and cissoid. They also knew the conic sections, and Apollonius ("the Great Geometer," as the Greeks called him) wrote a treatise on the subject in the third century B.C. Neither he nor his contemporaries had any idea of the use that Kepler would make of them toward establishing modern astronomy, the most imposing edifice that science has erected.

Cavalieri, of the lay order of Jesuates (*Jesuati*), who taught at Bologna, wrote a book on the infinitesimal in geometry. It was published in 1635, a few years before Newton was born. He, too, could not have foreseen the use that would be made of his work by the greater makers of the calculus in the generation following.

Léonard Euler and a number of his contemporaries in the eighteenth century wrote about the "imaginary number." It may have seemed a useless waste of time to most of those who knew him; but it led to some slight appreciation of the significance of such numbers a little later, to the inventions of Sir William Rowan Hamilton *(Quaternions)* and Hermann Günther Grassmann *(Ausdehnungslehre)*, who applied their methods to physics. Still later these same imaginaries became even more real in the invention of vector analysis by J. Willard Gibbs in the last two decades of the nineteenth century. The work of Gibbs has proved of great value in the study of physical problems, and civilization has bettered by the result.

These are a few of the illustrations of cases in which apparently useless studies of special branches of mathematics have, years or centuries later, proved to be of great service to humanity and to be contributions to our civilization.

Contributions to the fine arts. The contributions of Mathematics to the Fine Arts are chiefly in three lines: (1) the theory of perspective, (2) the beauties of proportion, and (3) the theory of music.

Of these, the first has occupied the attention of the Western world (Europe and America) much more than the Eastern (China, India, Japan, Iran, and Iraq). Euclid wrote about it in his *Perspectiva*; Vitruvius mentions it in his *De architectura*; in the Middle Ages it was studied by Roger Bacon and numerous other scholars. As a mathematical subject applied to the graphic arts, however, it may be said to have had its beginning in the fifteenth century, and in Italy. By the middle of the sixteenth century it was taught in all the important schools of painting of Europe. Sometimes, as in the works of Canaletto, the subjects of the paintings are so selected as to make perspective seem to dominate the picture. In any case, this contribution of Mathematics is pronounced in the works of the European painters, and is lacking in much of the work of the Orient.

As to the second, the use of proportion in the arts, there is a considerable literature upon the beauties of symmetry, axial and central, beauties found in nature as well as in art. There is also literature upon the beauties of lines divided in extreme and mean ratio, known by the modern name of "Golden Section." A line AB is said to be divided at P in Golden Section when $AB : AP = AP : PB$. If a line is eight inches long, it is approximately divided in Golden Section when the two parts are five inches and three inches.

Nature seems to build on the basis of axial symmetry, as in most leaves; circular symmetry, as in the cross sections of many fruits or in certain crystals (e.g., snow, quartz); or the Golden Section, that is, about 3 to 5 as in the arrangement of certain branches of trees, as in that of leaves on a vine or the branches of a fern leaf. Moreover, in art the approximate 3 : 5 division is often used unconsciously as well as consciously. This is seen in the dimensions of classical buildings like the Parthenon, of flags, of book pages, or of wall decorations.

In the line of beauty, therefore, Mathematics makes a constant contribution to the most pleasing phases of our civilizations.[6]

Those who wish for a scientific discussion of the mathematical

[6] For an extended exposition of this phase of the subject see Bowers, Julian, "Dynamic Symmetry," *Scripta Mathematica*, I, 236-245, 309-314.

theory of aesthetics should read Professor George D. Birkhoff's work, *Aesthetic Measure* (Harvard University Press, 1933). The author has there shown the close dependence of mathematics and the fine arts, each upon the other, and has done this with an unusually stable background of the former and a genuine appreciation of the best that is in the latter.[7]

Looking backward. In considering any social, political, ecclesiastical, scientific, or educational problem, it is interesting and often helpful to look backward and see what our ancestors had to say upon the subject. In this way we find how the world has progressed in language, in methods of presenting an argument, in education, in the needs of people, in the various branches of knowledge, and in what we call our civilization. It is for such reasons that it is worth while to consider what were thought, nearly four centuries ago, to be some of the contributions of such a simple subject as elementary arithmetic to the civilization of the time.

The following quotation is from one of the most celebrated textbooks that the English ever produced, *The Grovnd of Artes,* written about 1540 by Robert Recorde. It was composed like a catechism, a dialogue between the pupil ("Scholar") and teacher ("Master"), and began with "The declaration of the profit of Arithmeticke." The book went through many editions and revisions, and the following excerpt is from the one of 1646, a century after the first. This edition is chosen because it contains additions and revisions by such teachers as John Dee and John Mellis and may therefore be said to represent the best educational views of the sixteenth and seventeenth centuries. It is here set forth with the original spelling, punctuation, and sentence construction.[8]

Scholar. I perceive by your former words, that Astronomy and Geometry depend much on the help of numbring: but that other Sciences as Musick, Physick, Law, Grammer, and such like, have any help of Arithmetick, I perceive not.

Master. I may perceive your great Clerk-linesse by the ordering of your Sciences: but I will let that passe now, because it toucheth not the matter that I intend, and I will shew you how Arithmetick doth profit in all these somewhat grosly, according to your small understanding, omitting other reasons more substantiall.

[7] There is a review of the book in *Scripta Mathematica*, II, 55-59, by Professor D. N. Lehmer, who has entered into the spirit of the work with a thorough understanding of the treatment. The perusal of this review will assist in the reading of the rather technical book.

[8] From a longer quotation in the author's *Source Book in Mathematics*, 1929.

First (as you reckon them) Musick hath not onely great help of Arithmetick, but is made, and hath his perfectnesse of it: for all Musick standeth by number and proportion: And in Physick, beside the calculation of criticall dayes, with other things, which I omit, how can any man judge the pulse rightly, that is ignorant of the proportion of numbers?

And so for the Law, it is plain, that the man that is ignorant of Arithmetick, is neither meet to be a Judge, neither an Advocate, nor yet a Proctor. For how can hee well understand another mans cause, appertaining to distribution of goods, or other debts, or of summes of money, if he be ignorant of Arithmetick? This oftentimes causeth right to bee hindered, when the Judge either delighteth not to hear of a matter that hee perceiveth not, or cannot judge for lack of understanding: this commeth by ignorance of Arithmetick.

Now, as for Grammer, me thinketh you would not doubt in what it needeth number, sith you have learned that Nouns of all sorts, Pronouns, Verbs, and Participles are distinct diversly by numbers: besides the variety of Nouns of Numbers, and Adverbs. And if you take away number from Grammer, then is all the quantity of Syllables lost. And many other ways doth number help Grammer. Whereby were all kindes of Meeters found and made? was it not by number?

But how needfull Arithmetick is to all parts of Philosophy, they may soon see, that do read either Aristotle, Plato, or any other Philosophers writings. For all their examples almost, and their probations, depend of Arithmetick. It is the saying of Aristotle, that hee that is ignorant of Arithmetick, is meet for no Science. And Plato his Master wrote a little sentence over his Schoolhouse door, Let none enter in hither (quoth he) that is ignorant of Geometry. Seeing hee would have all his Scholars expert in Geometry, much rather he would the same in Arithmetick, without which Geometry cannot stand.

And how needfull Arithmetick is to Divinity, it appeareth, seeing so many Doctors gather so great mysteries out of number, and so much do write of it. And if I should go about to write all the commodities of Arithmetick in civill acts, as in governance of Common-weales in time of peace, and in due provision & order of Armies, in time of war, for numbering of the Host, summing of their wages, provision of victuals, viewing of Artillery, with other Armour; beside the cunningest point of all, for casting of ground, for encamping of men, with such other like: And how many wayes also Arithmetick is conducible for all private Weales, of Lords and all Possessioners, of Merchants, and all other occupiers, and generally for all estates of men, besides Auditors, Treasurers, Receivers, Stewards, Bailiffes, and such like, whose Offices without Arithmetick are nothing: if I should (I say) particularly repeat all such commodities of the noble Science of Arithmetick, it were enough to make a very great book.

The educational problem. The fact that Mathematics is the rock upon which the arts and sciences of the world rest does not mean that it must be taught extensively to boys or girls in our schools. Children are not supposed to be versed in the treatment of smallpox or epilepsy just because the physician must have this knowledge; but they should know something about first aid, hygiene, and the value of vaccination. They need not learn the mineral resources of Meshed, or where this place is located, but it is desirable to know something about the general products of the leading parts of the world. Similarly, a pupil in the high school need not know what the calculus is, although it is quite possible to give him some idea of the subject; but in the freshman year at college it is desirable that he should come in contact with a subject of such far-reaching importance. Speaking more generally, it is desirable that every well-educated person should have some idea of the bases of human knowledge. Needless to say, the school can simply open the doors of this knowledge. To know the significance of history; to come in contact with certain models of the best literature, music, and the other fine arts; to know what the sciences mean in the intellectual and physical life of today; to realize our duties to society; if possible to know something of another language than our vernacular; and to understand the significance of mathematics as one of the most important bases of all scientific work—all this should be a part of the training of our people. This is not a plea for making expert mathematicians, physicists, musicians, botanists, or weavers at a loom; it is a plea for giving a general idea of the significance of the great branches of human knowledge. I often recall the remark made to me years ago by a scholar who came to the United States and became one of our valued citizens. He said, "Almost the first thing that impressed me most was the apparent fact that American educators are so uneducated." Were our schools to blame then? What of the present?

The Seven Lamps of Mathematics. As a further summary of the contributions which mathematics has made, let me refer to The Seven Lamps of Mathematics which, it seems to me, have served to illuminate civilization through the ages.[9]

The first of these lamps is the *lampas utilitatis*, for we cannot convey mathematics to the great mass of people unless we first

[9] The following quoted paragraphs are, with slight alterations, from an address by the author published in *The Mathematics Teacher*, XIX, 279, 280. See also "The Call of Mathematics," *ibid.*, 282-290, and "The Poetry of Mathematics," *ibid.*, 291-296.

dwell upon the utility of the subject and imagine what would happen to the world if every trace of mathematics and of mathematical knowledge were blotted out tonight.

The second is the *lampas decoris*, the lamp of beauty; because if we are to teach mathematics at all, real success is not possible unless we know that the subject is beautiful as well as useful and can pass this knowledge on to our pupils. Mere utility of the moment without any feeling of beauty, becomes a hopeless bit of drudgery, a condition which leads to stagnation.

The third has been the *lampas imaginationis*, which has always seemed to me especially appropriate in referring to a medieval cathedral in which we set our lamps, and which seems equally so in respect to our chosen science; for what would mathematics have amounted to without the imagination of its devotees, its giants and their followers? There was never a discovery made without the urge of imagination, of imagination which broke the roadway through the forest in order that cold logic might follow.

The fourth is the *lampas poesis*, the lamp of poetry; because if one does not feel the poetry in mathematics, he may as well cease teaching the science. What, after all, is mathematics but the poetry of the mind, and what is poetry but the mathematics of the heart?

The fifth lamp we all seek to light is the *lampas mysterii*. This it is that reveals to us one of the great charms of the science—that in working in the domain of mathematics, we are surrounded by clouds, and success drives back these clouds a little way, and a discovery is made; then someone makes another discovery and drives them back a little more; and at rare intervals in time a Newton comes and drives them back what seems a long, long way, and an Einstein pushes them still farther—and still there is the surrounding mist of mystery. It is a great experience, this piercing the clouds; but try as we may, there is still the mist about us.

The next is the *lampas infinitatis*, the lamp of the infinite. A writer not long ago, in a verse which appeared in one of our magazines, spoke of mathematics as the science which lassoes the flying stars. It means much to have played even a little around the outskirts of "the science venerable," to have seen how it reveals something of our own position in the macrocosmos, and to see what infinitesimal things we seem when we look at ourselves in the light that mathematics sheds upon this great cosmos. The other day a mathematician took a certain measurement, and it was by no means one of the greatest of our time. He found that one of the other universes about us was six quintillion miles away, one million light years. The greatest speed that we can obtain mechanically by any present means is the speed of a rifle bullet, which may go half a mile in a second—a velocity so great that we can hardly imagine its possibility. If we ask how long would it take the rifle bullet to reach that other universe, even our very elementary mathematics gives us the answer—it would take three hundred eighty billion years. Truly it is "the science that lassoes the stars."

And the seventh of the lights is the *lampas religionis*. We may wonder if such a candle burns and sheds its light, but I have an idea that we all feel that, while a mathematician may not necessarily be a very religious man, on the other hand no man can appreciate religion to the full unless he has to assist him some knowledge of the great field which mathematics opens to his vision. Mathematics may not make any man more religious, but if he is religiously inclined it makes him see the grandeur of religion as nothing else can.[10]

[10] See an address by the author entitled "Religio Mathematica," *American Mathematical Monthly*, October, 1921; and the *Teachers College Record*, November, 1921.

THE CONTRIBUTION OF MATHEMATICS TO EDUCATION*

By SIR CYRIL ASHFORD

Late Headmaster of the Royal Naval College, Dartmouth

A SCHOOLBOY wrestling with the elements of arithmetic or algebra cannot be expected to have any grasp at all of what mathematics really is. But any educated man who has, as it were, stood back and looked at mathematics as a whole will agree that it stands almost uniquely as a stupendous structure, a monument to the human intellect. It deals with abstractions in a perfectly ordered and logical manner; it is in no wise concerned with human values, and it scorns contact with material things. You may supply it with postulates or assumptions which are startling or absurd from our limited human point of view, and it will duly produce the conclusions that logically follow. Indeed, that is the normal procedure in, for instance, the geometry of Lobatchewsky, the four-dimensional space-time continuum essential in Einstein's work, or the still more unimaginable five- or six-dimensional space which is needed to reconcile the quantum and the classical mechanics.

Although mathematics holds itself aloof from practical affairs, it supplies the practical man with the tools which are absolutely essential to his everyday work—ranging from the village carpenter to Signor Marconi or the designers of aeroplanes, from the cashier in a teashop to the Astronomer Royal.

This edifice has been built up by the devoted labors and genius of a long series of investigators, and it is now so vast that it would require most of the working lifetime of a picked brain to comprehend the whole of it; probably no living mathematician would claim such complete comprehension, since he would have been attracted away into research in the part that specially interested him before he had mastered all that has been achieved in other directions.

* Reprinted by permission of The Macmillan Company from Harrow Lectures on Education.

Granting all this, the questions that probably arise in the minds of many educated men may well be something like the following: Is it advisable to force all boys to essay these very abstruse and alien modes of thought, as part of their training for life? It is doubtless magnificent, but is it education? Is not mathematics comparable, so far as the ordinary man is concerned, to an enormously elaborated game of chess, which may safely be left to chess-masters with unpronounceable names? Is it not so entirely unrelated to everyday life and thought and human intercourse that it may be relegated to the fourth dimension with which it is prepared to deal? A certain, very limited amount of arithmetic, too simple to be called mathematics, is undoubtedly essential to the conduct of ordinary life—why not let boys be drilled in this, and leave the rest to such as enjoy that kind of game, who will continue to provide another set of abnormal people, the theoretical and practical scientists and engineers, with the machinery needed for their work?

It is probable that you cannot as yet dictate the policy or curriculum of any of the great public schools; but sooner or later, as tutor, house tutor, housemaster, or headmaster, nearly every public schoolmaster becomes concerned with all parts of that curriculum and its reaction on the boys who are his especial charge. So it is of vital importance that if you have such ideas as I have suggested, their expression should not be inhibited because of a fear that they may seem narrow minded or heretical, to become in the jargon of psychologists a "complex," with evil results to yourselves and those who live with you.

Let us therefore face the question with complete frankness, and so far as possible discard all prejudices. It is easy for me to do so, since I have passed the stage when it matters what anyone says or thinks of me—and it is fortunately easy for you, since you have only to listen and are not necessarily called on to express dissent or concurrence.

If you look into the history of education you will see that at different times these questions have been answered in different ways. Down to about a hundred years ago, all European peoples with the exception of the Greeks effectively voted against mathematics as an educational subject. A few schools, such as Christ's Hospital and the Mathematical School at Rochester in Newton's time, taught a good deal of mathematics for technical purposes. But the absence of such teaching in ordinary schools was not due to contemporary ignorance of the subject. For higher mathematics, as we should now rate it, was pursued with great eagerness and success at the universities, in England and on the continent. Those in charge of school curricula, if they thought about it at all, must have frankly condemned it as a subject for boys.

Do not fear the worst if I admit that I am going to quote from my own experience of the seventies and eighties of the last century; this is not a Speech Day address in glorification of the past. On the contrary, I am prepared to assert that, under the conditions prevailing at that date, it would be difficult to refute the objections which I have put into the mouth of the Devil's Advocate. But I hope to be able to convince you that his general conclusions are sound only because the school teaching of mathematics was then designed on the wrong principles, not because mathematics is an unsuitable subject for the education of schoolboys.

Let us first inquire how much validity there is in his complaint that mathematics employs abstruse and alien modes of thought, unrelated to everyday life. A survey of the process by which a boy reaches the multiplication table will serve our purpose, in this and other respects.

The normal course of imparting the very earliest stages of arithmetic—it would be absurd to call them the simplest stages, though they are the most elementary—consists in making the child deal with groups of matches, dots, inches, or any sets of similar objects, with his own hands. He learns to use the conventional names for the number of similar objects in the group, and by adding and subtracting other groups he learns to work various little oral sums in addition and subtraction, each result being obtained by direct experiment in one concrete instance. By changing from matches to the same numbers of dots or counters, and observing the similarity of the results, he gets the first inkling of the idea of generalization. Later, he reaches the stage of visualizing the groups of matches, etc., and "working in his head." After a great deal of practice of this kind he may attain to some notion of pure number as an abstract idea—but it is a long step from the experimental result, "If we add 2 matches to 2 matches and add another 2 matches to them, we get 6 matches," to the statement, "three two's are six." Most little boys will not really comprehend the latter phrase until they have done very many experiments with a variety of numbers —for abstraction is a more difficult process than induction.

The common practice is to deal at this stage only with small numbers, up to say 9: the shorthand for the names or so-called Arabic symbols (which are really Hindu) can of course be taught at any time when it seems desirable to begin writing down the problems. I suppose most children associate small numbers with definite patterns, as on playing cards; such numbers are not for them purely abstract symbols. But for larger numbers, such as 144, patterns are useless. So boys are introduced to the device of making up bundles of 10, and to the notation on the decimal system, with its place value for the various digits. The idea of the series formed by successive natural numbers is also rubbed into them, so that they can get a notion of the relative magnitude of largish numbers by their positions in this series, by the times it would take to count up to them from one, and from the length of those numbers of units of length put end to end. So the larger numbers probably represent rather more purely abstract quantities to them than the smaller, and their relative magnitudes are a little less definite. But persistent return to concrete examples enables them to grasp the meaning of the multiplication table up to 12×12. They then have to learn every individual item of it by heart, and practice themselves in it until it becomes part of their mental equipment, unfailingly ready to be produced automatically whenever needed.

In this long and tedious process, the numbers have doubtless become to them pure numbers, and perhaps have ceased to possess much meaning unless they are called back from their shadowy existence in order to help in solving some problem dealing with the concrete. I am not sure that this does not happen even with mathematicians.

It will be readily admitted that every member of a civilized community must attain to a mastery of at least this amount of power of dealing with abstract symbols, and that he can most easily attain it by this process. So some at least of the modes of thought that are of fundamental importance in mathematics, though they may be abstruse, are by no means alien or unrelated to everyday life—except in their philosophical description they are not at all caviar to the general.

I selected this example partly because you have probably forgotten your own experience of it in your extreme youth. That in itself may be taken as a testimony to the success with which you were taught—I have nothing but praise for the skill and patience of our nursery governesses. But when we go on to consider the mathematical work in the schools, we find a very different state of things.

Since the broad outlines of the method of guiding a boy's first steps in generalization and abstraction have probably changed little in the last fifty years, we can assume that in the 1880's boys went to their preparatory schools in much the same mental state as in the 1920's.

But those earlier generations were then treated in what we should now consider a preposterous manner, in leading them on in arithmetic and starting them in geometry and algebra. It was practically assumed that they had once and for all mastered the idea of abstraction, that they had no further need of inductive processes, and that the aim was to treat everything henceforth by deductive logic. Merely as an instance of this point of view, which, however, throws a flood of light on it, I would remind you that textbooks on arithmetic and algebra nearly always gave blocks of examples of manipulation of symbols, to be worked out on the lines of a specimen in the text. When a boy could do so and obtain the answer at the end of the book, he could pass on to the next block. There was very little "explanation" given in the text—that was supposed to be done by the master, but it is safe to say that it was not always enlightening. There were also blocks of so-called "Problems" which may be described as the application of symbolic methods to concrete questions; but the average boy was so hopelessly out of his mental depth that it was quite impossible for him to come back from the abstract to the concrete—so he was "excused problems."

On the whole the treatment of geometry was even worse. It reduced itself, for the average boy, to learning by heart the axioms, postulates, and theorems of Euclid, with little or no hope of understanding what it was all about. There were no riders in Euclid's text; if Todhunter added some, they were for the mathematically elect. It is easy to see now why the schoolmasters failed where the nursery governesses succeeded. The former came down to the schools from the universities with their minds full of their special subject, and proceeded to expound it as it had been expounded to them, without much reference to the powers of the learner—while the governess knew little of the subject but possessed a great deal of sympathy and insight into what is now called the child mind. It must also be remembered that in those days there were no Courses for Junior Public Schoolmasters or Training Colleges for Elementary School Teachers or Professors of Education at the newer universities—few or none of such societies as the Mathematical Association, the Science Masters' Association, the Modern Languages Association, the English Association, or the Classical Association (although the Association for the Improvement of Geometrical Teaching was an early forerunner) and few or no clearing houses for educational ideas such as the *Journal of Education.* A public schoolmaster was to a great extent a law unto himself. On joining a school he was given a form or division to cope with single-handed, with not much support in the maintenance of discipline and no guidance at all in his teaching. If he failed he left, and if he succeeded he became entrenched in an impregnable position with every inducement to believe in his own powers as a teacher and in the perfection of his methods of teaching. They were indeed "the good old days" for the master, but perhaps not for the boys, or for the art of education.

I honestly do not think I have exaggerated the badness of the school teaching of mathematics at that time. If the picture is even reasonably true, what impression do you think the situation will have on the minds of the majority of boys who were forced through the process during the whole of their time at their preparatory school and public schools, until they could manage to drop the farce of learning "mathematics"? It is quite obvious that even if they had successfully taken the earliest steps in the comprehension of the abstract, and the meaning and use of symbols, their subsequent experience must have been like a fog descending on them, which, like all fogs, caused blindness and exasperation. And they were the men who have now attained to mature years and correspondingly authoritative positions—whose opinions carry very great weight. I submit that it is asking more than is humanly possible if we now expect them to put away altogether from their minds the bitter memories of their youth, in regard to the "Contribution of Mathematics" to their education. Frankly, it is at least doubtful whether that contribution had much intellectual or utilitarian value, though as a training in some of the moral virtues there may have been something to be said for it.

A certain number of the victims struggled through, and by dint of laborious effort and natural aptitude at long last attained a clearer conception of what had been almost wilfully obscured—but I believe that their earlier struggles handicapped rather than profited them. Probably if they had postponed their start in mathematics until their minds had been matured by age or other learning, they would have gained in the end. This was effectively the case with the rapidly increasing number of men who devoted themselves to the study of science. They found that mathematics was for them an essential tool; so they had to set themselves to mastering so much of it as they required. They mostly began with what is now called "Practical Mathematics," and, after all, that is not a bad way in which to begin. When they were teaching themselves, with a definite object in view, and at an age when deductive reasoning and abstract ideas are congenial, they naturally made rapid progress and largely overcame the handicap of their misspent youth.

The direct outcome of this has been the separation of a definite stratum of educated Englishmen into two camps—one devoted to Humanities, the other to Science, using that term fairly broadly. And it is largely a class distinction, for the former group came mostly from the older public schools; a few of the latter came from those schools, but most were recruited from less socially distinguished schools. The consequent inferiority complex on the part of the followers of Science embittered their attacks on the "grand old fortifying classical curriculum," while its supporters could congratulate themselves on not being as those other men are, and tended to ascribe it to the curriculum they had enjoyed at school. It is this which makes the separation into two camps a serious matter. Such separation is inevitable to some extent, for the tastes and aptitudes of different men will always lead them to specialize in what may be broadly classed in one or other of these categories. But these tastes and aptitudes are not, I am convinced, statistically a function of birth or social status; where the evidence seems to suggest that they are, I believe it is due to the pressure of home or social environment. When they have had time to adjust themselves, each group will contain men of all social grades, and one cause of misunderstanding will have disappeared.

The end of this process is now fortunately in sight, and the places of men of that generation are gradually being taken by men who have not been warped in their youth. So we may pass on to a more cheerful topic, and look for the cause of the change which has taken place in school mathematical teaching.

About the beginning of this century—or some years earlier in a few schools—the physical sciences secured a real foothold in the public schools, with effects that have been revolutionary in many ways. The one we are chiefly concerned with at the moment is the effect it has had in modifying the teaching of mathematics. It now makes, for the first time as I think, a real contribution to the education of the ordinary, not specially mathematically minded, boy, and has widened the intellectual equipment of the future mathematician and scientist—a smaller matter, but by no means negligible.

The revolution in the school teaching of mathematics of course began many years before the end of the century—the A.I.G.T. did great work in the seventies, but it lacked driving force. This driving force was supplied, in my belief, by the astonishing growth and importance of science in the outside world.

Physics and chemistry were reluctantly admitted as school subjects and were treated at first very much as a Cinderella. They were undoubtedly a nuisance; and consent to their introduction was only granted when the external pressure became too great to resist. Once they were made available, however, a surprisingly large number of boys found them decidedly congenial to their temperament; and these boys were not solely the dullards, as classical masters used to assert. So internal pressure was added to that from outside, and the extension became very rapid.

A reaction on the teaching of mathematics was inevitable, if only because of the immediate demand that physics made on the mathematical equipment of boys, which had previously been of value to them solely for passing examinations. But, standing alone, that would not have sufficed to produce any profound modification in the methods of teaching mathematics. What was, I think, the prime cause of the revolution that occurred at this time was the demonstration to the mathematical masters of the suitability of the experimental method and inductive processes to the mental age of the boys. To anyone who had attempted to teach along the old lines of abstraction and deduction the response to the methods of science was a revelation; and although the bulkheads between departments in a school were, and unfortunately still are, fairly watertight, the effects on the boys were so obvious that they could not pass unnoticed. So with commendable breadth of mind the mathematicians on the staffs of many schools set themselves to adapt their teaching so as to take advantage of this new knowledge.

As the revolutionary ideas were born in the schools and not in the universities, and the changes were needed only in the schoolboy stage, it called for much hard work and persistence on the part of the reforming schoolmasters to get their views accepted by the authorities responsible for external examinations; but in the course of a comparatively few years it became universally admitted that the traditional conception of the proper way to present the earlier stages of mathematics to boys had to be revised. A rather chaotic period supervened, but something like agreement has now been reached on the broader issues. This agreement can be put briefly (and of course adequately) as the application, as a boy begins any fresh branch of mathematics, of the methods which I sketched as being traditional in the teaching of the multiplication table.

For instance, in geometry he is given an extended course of practical work, using a graduated ruler, protractor, etc., which were taboo to Euclid, until he has a working knowledge of the material from which the concepts of formal geometry are abstracted. Through this practical work he acquires enough general notion of the subject matter of the postulates and axioms of Euclid to make it quite unnecessary to formulate them in precise terms for the earlier course in formal geometry; such formulation admittedly does not help a boy to comprehend them. This earlier course is to be regarded merely as an attempt to regularize his experiments and intuitions, and not an attempt to practice him in strict logic. But it serves as an admirable step towards a later, more rigorous treatment of the same material if he perseveres with the subject.

In the same way, algebra is introduced as generalized arithmetic; trigonometry is prefaced by a great deal of numerical trigonometry; statics and dynamics are taught experimentally at first; and so on. All these changes are so recent that I need not inflict on you a detailed description and analysis of the procedure now adopted.

Stating the position broadly, I believe it is safe to say that mathematics is now for the first time taught to boys in such a way that all but the very stupidest can without undue effort understand what they are doing and, in a general way, why they are doing it. Methods have been adopted which are suited to their mental age. The subject no longer stands in splendid isolation, but is intimately connected with another school subject, both depending on the other for mutual help. Science asks mathematics for essential machinery; mathematics ask science for essential material; and they employ at this stage largely the same methods.

This brings me to what is rather delicate ground. I am going to claim the privilege of age and venture to give those of you who are engaged in teaching mathematics a little good advice. This revolution in the teaching of elementary mathematics must, I imagine, be credited to your seniors. It is not a complete and final reformation, incapable of further improvement, but rather a change of attitude towards a very difficult and complex problem. It would therefore be fatal if you folded your hands in smug self-complacency. I do not for a moment imagine that this is likely—so I am tempted to suggest to you one way in which, if I have correctly analyzed the cause of the recent improvements, you can profitably exert yourselves to carry on the good work. This is, in brief, to get in touch at every possible point, and to keep in touch, with the physics teaching that is going on in your schools.

I know quite well that this is not easy. The more highly organized a school becomes, the greater is the pressure to keep a man in his own department. I confess that, although throughout more than twenty years as a headmaster I set myself to put this excellent principle into practice and not only facilitate but almost compel men to work outside the narrow limits of their own specialty, the extent to which I succeeded fell far short of my hopes. Human nature is very strong, and persistent in its influence; a rut is a comparatively pleasant track. The alternatives are, on one side, smooth running with the maximum of immediate efficiency, on the other, troublesome dislocation of habits and arrangements for the staff and the provision of less expert teaching for the boys; in these circumstances it is very tempting to keep masters doing their own special jobs. But I am convinced that an occasional change of pulpits between mathematicians and physicists is of immense value in the long run, and that the benefits far outweigh any present discomforts and loss of efficiency.

If this can be arranged officially, so much the better; if not, much can be done on the initiative of individual masters who are prepared to sacrifice some of their scanty leisure, and by exercise of diplomacy get their colleagues' permission to act as an assistant demonstrator (unpaid) in a physical laboratory, or swell the audience in a lecture room. This is not so difficult as it may appear, if the physicist can be made to realize the advantage of seeing with his own eyes how well mathematics is now taught, and how little grounds there are for his complaint of boys' ignorance of practical mathematics and lack of skill in manipulation of symbols.

In urging greater co-operation between mathematicians and physicists I may be suspected of forgetting the mathematical laboratories which have been established at some schools. These are of course far better than nothing. In a school where no physics is taught, they would be invaluable, but I doubt whether such a school now exists. Under modern conditions they are open to serious criticism; for example, to a scientist there is a sense of unreality in using a Boyle's Law apparatus with the primary aim of plotting a rectangular hyperbola, or in investigating the connection between the length and period of a simple pendulum because it furnishes an easy example in algebra. But to my mind one of the most weighty objections to them is that they represent a weak submission to the forces opposing co-operation. They are devices to make life easy for masters rather than profitable for boys. The boy will of necessity pass to and fro between the physical laboratory and the mathematical classroom, getting something of the spirit peculiar to each. It is far better when the master does so too, learning when he is outside his own domain, and on his return using his new knowledge in still further improving the methods for teaching his own subject. And this is not the only, or the greatest, benefit that he will reap from his travels abroad. For the frontier between mathematics and physics is not easy to delimit; there is a great deal of interpenetration and overlap; it is almost true to say that it is only in the school stage that overzealous organizers have succeeded in marking out a sharp boundary between them. Now, one of the contributions to education which is rightly claimed for mathematics is what is called "outlook on the modern world." There is no outlook from a watertight compartment; to get one we must ascend to the bridge, or some deck above which the bulkheads do not extend. So if we want a boy to go into the world with a desire for a wide outlook and some practice in obtaining it, we should provide him with masters who themselves have not only the right to pass from one compartment to the next, as the boy must do, but also some of the faculty of co-ordinating the activities in the separate compartments that are usually found among those assembled on the bridge.

Since I have, in the last few words, mentioned one definite contribution to education, it may be as well to apologize for not attempting to deal directly with what may seem the proper subject matter of this paper, the actual contributions which can be expected from mathematics when properly taught. My excuse for not doing so is twofold. In the first place it has already been done in countless books, most of which are readily available—notably in the Report of the Mathematical Association Committee on the Teaching of Mathematics in Public and Secondary Schools. I certainly could not hope to improve on that.

But my main reason is that the history of the changes that have rendered these contributions actual instead of theoretical furnishes a most striking illustration of the supreme importance in education of the reaction between various subjects, and of the disastrous results of isolation. I know of no other branch in which the evidence is so direct and so conclusive—and at a conference of this kind, which is composed of all the talents, it should be of greater interest and profit to consider this aspect of the matter than to attempt to confirm the faith of those who already believe, or convert the doubters, or inform the ignorant by special pleading directed to establishing the benefits derived from the inclusion of mathematics in general education.

I am very conscious of the fact that I run a certain amount of risk in making this choice of topic. No history is so little taught as recent history—the expert historian usually refuses to touch it on the grounds that it is still too near to be in perspective, that it is too imperfectly studied for a sound judgment to be possible, and that it involves people who are still alive. So the amateur has to rush in, as I have done, and risk being dubbed a fool for his pains.

And that brings me to my last point. I have drawn a gloomy picture of school mathematics in the past. It would be disastrous if I left you with the impression that the picture is now anything but bright. I have no hesitation in saying that under present conditions there is full warrant for the claims put forward by mathematical experts, even for that extreme one at which so many men of an older generation have laughed, bitterly or derisively, according to their temperament, the claim that the average boy can actually enjoy his mathematical work.

Since I have been taught that the end of a sentence is the position of greatest emphasis, it is on that cheerful note that I now end.

MATHEMATICS IN GENERAL EDUCATION

By W. LIETZMANN

Göttingen, Germany

The meaning of mathematics. Many persons associate mathematics with some geometric propositions or algebraic operations, like the theorem of Pythagoras, the binomial theorem, or the solution of quadratic equations; the more meticulous might think of differential and integral calculus, differential equations, or Fourier Series. Thus, those who have to do with mathematics outside of school life are an awe-inspiring minority. But simple operations with numbers, the development of the concept of percentage, the computation of interest, as well as the knowledge of any graphical or numerical conception of plane or solid geometric forms, belong to mathematics.

When practical experience embraces these phases, however, the feeling that one is dealing with a science disappears. The long division we study at present was a subject taught in universities in the sixteenth century. The function concept and graphical representation were given scientific meaning only during the last century, when they were practically applied in certain branches of technology and applied natural sciences. Only at the turn of the nineteenth century had the rising mathematical reform movement introduced and spread this concept in the schools; now there is no newspaper, no popular work dealing with facts represented by measurements, which does not employ graphical representation of empirical functions. Here also it is not quite obvious that mathematics is involved.

When we consider mathematics in this scope, at the expense of depth, its part in general education becomes quite apparent. These arguments do not exclude the education of the future research mathematicians or other groups of science students, such as the astronomer, the theoretical physicist, the research technologist, for whom higher mathematics is a necessary tool. But the sphere of the application of this science is much wider.

Vocations and Mathematics[1]

Mathematical applications. It is very important to make clear, once more, how great the field of mathematical applications is, when mathematics is considered in a broad sense. We shall consider groups of vocations and we shall inquire how useful and necessary mathematical abilities are for them. We shall not consider the mathematical subject matter, such as the knowledge of certain definitions or theorems, the mastery of certain algorisms and methods, as much as the general dependence on mathematics and the discovery of specific mathematical abilities.

The merchant. The merchant in the broad sense, be he storekeeper or wholesaler, farmer or shipper, must deal with numbers and measures of various kinds, with plane and solid objects, with measures of weight and time, perhaps also with speeds, forces, and measures of work. For all these he must possess a quantity-feeling, which will enable him to determine prices quickly and correctly; to calculate them, and on this basis organize his business. He who cannot do this is himself the loser, or he is the cause of misfortune, especially when heavy taxes, ruinous duties, or emergency demands are imposed.

The artisan. In the case of the artisan, the ability to calculate is important, because very often he is at the same time a business-man. But the mathematics of bricklayers, plasterers, painters, carpenters, tinsmiths, of automobile-mechanics especially, and of electricians who are also engineers, is a mixture of arithmetic and plane and solid geometry. The method of thinking from a mental image through a drawing on paper to the ready-made product requires the ability of space perception. The young man, for example, who does not master the various kinds of wood-combinations is useless as a carpenter; the mechanic who cannot read a blueprint is equally incompetent. The examinations for licensing the various trades involve an amazing extent of mathematical problems.

The artist. Many artists proclaim themselves enemies of mathematics. But the wildest expressionist, who professes to be the despiser of all perspectives, must look out for the true forms, especially

[1] This subject was discussed by the writer on another occasion in an article: "Die geistige Haltung des Mathematikers, Vererbung oder Erziehung?" *Zeitschrift für mathematischen und naturwissenschaftlichen Unterricht,* 66 (1935), pp. 369 ff., Teubner, Leipzig.

when he desires that the observer understand him. The plastic artist must not only be able to think in spatial terms, and represent in conformity with them; as a relief artist he must also master geometrical laws. That the architect not only must be a space-geometer but also must be familiar with certain number variations has been shown us recently by one of the most successful architects of our times, Th. Fischer.[2]

In primitive art, as in the highly developed ornament[3] and in hand-made tools and objects for the home, there is apparent a geometric "feeling" which is theoretically based on the group concept. The symbolic drawings of various peoples are based on definite knowledge of form which has been deciphered only in recent times.[4]

The scientist. In the vocational group of scientists, particularly the scientifically trained technicians, a vast range of mathematical skills is necessary. To this field belong not only physicists, chemists, astronomers, engineers, and technicians in all fields of endeavor, but also biologists, geologists, geographers, physiologists, psychologists, and specialists in economical sciences. Representatives of history, philosophy, and allied branches proclaim the nonmathematical nature of these sciences; yet chronology is their essential tool. While some judicious members of the medical profession and jurists appreciate the value of mathematical training, many of the so-called *Geisteswissenschaftler* (representatives of the belles-lettres, humanities, and the like) deny it and thus contrast themselves, not only to all exact sciences but also to manual workers and tradesmen generally.

How deplorable, for example, is the rift among present-day philosophers arising from the fact that some are able to think mathematically and some are not. In the times of Plato, Descartes, Leibnitz, and Kant, there was no such difference.

The citizen. Even if the few remarks concerning the mathematical needs of the various vocations have not completely indicated the breadth of the significance of mathematics, we must recognize that

[2] Fischer, Th., *Zwei Vorträge über Proportionen.* Oldenburg, München, 1934.

[3] Lietzmann, W., *Mathematik und Bildende Kunst*, Hirt, Breslau, 1931. Also by this author: "Geometrie und Prähistorie," *Isis*, 20 (1934), pp. 434 ff. "Geometrie und Urgeschichte," *Zeitschrift für mathematischen und naturwissenschaftlichen Unterricht*, 65 (1934), pp. 313 ff.

[4] The attention of the reader is called to the publications of the American Museum of Natural History in New York: Kroeber, A. L., *Basketry Designs of the Mission Indians* (1926); Wissler, C., *Indian Beadwork* (1927); Mead, Ch. W., *Peruvian Art* (1929).

the citizen in general, regardless of his vocation, is dependent on mathematics. He must balance his budget—make adjustment between his income and expenses; he must pay his bills, compute his taxes, evaluate life and fire insurance, and daily make decisions on the basis of comparative measures. As a citizen of a community or of a state he participates in community life and should interest himself in the conduct of the business of his immediate community and of the nation. Understanding of the business of the state, of commerce and trade, depends on mathematical abilities.

The individual who is interested in sports utilizes sports-mathematics. Mathematical abilities likewise contribute in large measure to successful participation in the so-called mental games, such as chess, cards, and puzzles. Indeed, mathematical thinking is a help which cannot be overestimated.

Space Perception

Geometrical concepts. Let us try to examine in detail what we call mathematical thinking and thus determine the educational value of mathematics. We begin with the development of space perception, a very obvious function in every mathematical study. We say very obvious; yet for quite a long time it was neglected. There were many teachers who thought, "Everyone can see; what is there to teach?" If models of simple mathematical forms were introduced, or better, if these were made by the pupils themselves, then these same teachers would say with an air of superiority: "We won't bring up workers for some cardboard factory, but mathematicians." In contrast to this, without presenting any objections to it at present, we say definitely: The teaching of space perception—when within the limits of his inherited capacities the pupil will otherwise be unable to comprehend relationships—is not only possible but is also necessary.

Everybody at some time or other must have a definite knowledge of how to recognize triangles and quadrilaterals as well as line configurations, to discover parallelisms and symmetries, especially when he comes in contact with them, and then proceed to the theoretical point of view. Certainly the road from actuality to the theoretical concept is the best.

In the case of such a conceptual development of geometric forms it is oftentimes necessary to simplify the "given" facts, but this abstraction offers the greatest difficulty on one side, while on the other, in the fixation in memory of contour lines (like land plans and maps), it represents (offers) simplifications.

For the selection of the content and method of mathematical study it therefore follows that the continuous interchange between abstract geometric forms and concrete facts is very important whenever a new geometrical concept is introduced.

Space geometry. Man lives in a world of three dimensions and not in Flatland. When, therefore, we go from the theoretical geometry of plane figures to spatial figures, and especially when we proceed from closed figures such as lines or bodies with a finite content, bounded planes, lengths, or other limited (in size) figures, such a step has a fundamental psychological meaning. Naturally, while we exist (live) in space, we should not be content to remain in the domain of the plane when we teach space perception.

We shall state three stages in the development of this phase of teaching.

1. In plane geometry the figures with which we work should be connected with spatial (three-dimensional) figures; thus triangles, quadrilaterals, and circles, for example, should be considered as plane sections of pyramids, prisms, and spheres.

2. We should enrich the study of plane figures and of similarity with numerical work and with drawings of simple figures of space, such as pyramids, prisms, cylinders, cones, and the like.

3. The relation between straight lines and planes in space, especially their graphical treatment, should be studied, according to the methods of descriptive geometry—by means of vertical and horizontal projections.

Space imagination. In all this we should not forget the following: Whenever we apply mathematics to some problem there is a beginning and an ending, but between these two there predominates the purely mathematical problem. The builder, the mining engineer, the landscape artist, the theater director, must first create his objects in terms of space imagination, before creating the drawing, and then he must turn it into reality, and whatever he has thus created he must afterwards submit to proof.

Not only is space imagination important for the creative individual, but it is also necessary for the understanding of events, whether presented in a novel, a drama, a description of exploration, a historical account, or in verbal descriptions of astronomical, geographical, and other phenomena.

Whoever is confronted with the theories of the structure of the macrocosmos of the universe or the microcosmos of the atom is immediately confronted with the difficulties of measuring magnitudes which are either uncommonly very large or very small.

Interpretation of space representations. Unfortunately, we must often resort to two-dimensional representation of spatial objects. Should such representation be true visual fixations, then the picture is easily understood by means of the central perspective. Paintings and photographs are common examples of such representations. Projections (*Schrägriss*), however, such as used by Japanese and Chinese and others, in their paintings, represent to some extent "visual" (*anschauliche*) pictures.

Artists and laymen have objected to the idea that the knowledge of fundamentals of geometry is of use to painting. They protest that it is an obstacle to the artistic feeling. In reply, we may state that the first and proper selection of the point of view and distance of the observer of a painting enables him to comprehend (feel) the representation better. In photography, for example, especially when enlargement is desired, proper location of the object in relation to the lense of the camera is important so that the plane picture may give a spatial effect.

In engineering we prefer, to the central-perspective representation, the representation that employs the vertical, horizontal, and base projections. Though more difficult for visualization, it is more helpful in the representation of size. During the last few years, in order to make the presentation and teaching of this subject easier in German schools, an orthogonal single blackboard method has been introduced.[5] It is connected by means of introduction of perpendicular projection and dimension-measurements with the making of maps and drawings. At any rate such topics should not be excluded from descriptive geometry in the schools on the ground that it has only to do with certain specific technical processes of certain trades. Any layman who has occasion to draw inferences concerning spatial shape of cross sections, or microscopic sections, or geological profiles, must understand and utilize horizontal and vertical projections.

[5] This is found now in all mathematics textbooks for German higher schools. These were introduced by G. Scheffers and W. Kramer in *Leitfaden der darstellenden und räumlichen Geometrie,* Quelle and Meyer, Leipzig, 1924; also W. Kramer "Einführung in die darstellende Geometrie," Part I *Math. Phys. Bibl.,* 66, Teubner, Leipzig, 1926.

Descriptions of form. We usually underestimate the difficulties of describing forms, whether they are plane or three-dimensional forms, whether they are simple, perfect, and with no unnecessary superstructures. Description must be studied. Unfortunately, slovenly description is prevalent in everyday use. The election law of a certain country, for example, states that during the election a four-edged urn should be used. The author of this law naturally did not think of a tetrahedron, but of a rectangular parallelepiped which has eight edges, but he saw only four.

The difficulties increase when we proceed from simple to highly developed forms or to combinations of form-groups. Let the reader attempt to describe fully the familiar form of a violin! The ornaments of the Gothic, and Roman architecture, the baroque, and the rococo can be distinguished easily, but to describe their differences in words is not easy at all. Such description of forms must always begin with the simplest figures, and in this case mathematics is a tool that has no substitute.

Samples of trends. The conditions in the world around us are not static. Our ideas are constantly changing. Trends are revealed in all fields—in technology, in natural sciences, in the spiritual sciences.

The geometry of Euclid was *static*, and it did not satisfy the requirements of an education. The geometry of our times has become more and more kinematic; motions, rotations, reflections, and transformations are common to it. Thus modern geometry so necessary in certain professions was developed. Here again we may draw examples from the various vocations.

Geometrical propositions. Only the specialists require applications of the propositions which are in the foreground of a course in school geometry and which are related to geometrical magnitudes, straight lines, angles, surfaces, and volumes in making constructions or in computing numerical values. Shall we say, then, that such problems as these should be confined to special vocational schools: the system of propositions concerning triangles, quadrilaterals, and circles, the study of surfaces, the study of similarity, in elementary geometry, then trigonometry and solid geometry and some other advanced topics in geometry. I cannot agree with this point of view. Any one of these abilities may at some time be necessary in the solving of even a simple problem. Moreover, geometrical perception, especially in relation to common problems, is best taught in the course of geometry. The instruction should incorporate real applications from life and thus develop the abilities in space imagination discussed earlier.

Development of Number Concept

Number concept. We usually distinguish two types of men who work with numbers: One who understand numbers, or think in numbers, the other to whom number problems are a terror. The first type allows quite instinctively a free play of a number feeling when it has to do with series, magnitudes, quantities, which is completely absent in the case of the second. To this second group, stock exchange records, statistical records, and even business records are unpleasant. Yet they employ the words *much* and *little*, *small* and *large*, *long* and *short*, which are associated with numbers; without a comparative scale these terms would be meaningless and empty.

The purpose of mathematical education of a great portion of youth is to develop various applications when these are especially of practical use.

Number comparison (evaluation). As it is important to develop a skill in correct performance of numerical operation, so it is important to develop the ability of proper comparison of various numerical magnitudes. Until we have developed the ability of comparison (by means of many problems), we cannot clearly comprehend the various geometrical or physical magnitudes. Here the comparisons are not confined to lengths, angles, surfaces, but include also groups of individuals and physical magnitudes such as the horsepower of a machine, strength of a current, carrying capacity, and thousands of other comparative items.

The number feeling enables the individual to estimate in advance the approximate result for a mathematical operation. The teaching process should provide training in judging the accuracy of results. Computations can be performed on a calculating machine, but a thinking computer is necessary in the preliminary estimate and in the final check on results. The number feeling is a protection against blind, purposeless numerical operations.

Number phantasy. The layman often fails to appreciate the careful mathematical planning behind technical projects, inventions, and products. Every machine, every airplane, every ship, every bridge is constructed according to measured specifications to render certain performances, number of revolutions, speed, carrying capacity, and the like. All of these capacities can be numerically determined. The preliminary work, before the project takes material form, cannot be done without an eminent sense of number feeling and space imagination. First, the project is a product of imagination; then by means of slide rule or calculating machine and drawing board, the first forms of thought become reality, and only then the execution can be started.

The objection may be offered that in general schooling the technologist or the constructor and their work are not considered. In small industry and in daily life, however, there are similar problems and applications. Various requirements and specifications are frequently expressed in terms of percentages. And here also the developed number feeling makes possible approximations in practical situations.

Functional thinking. Daily experience and scientific observation, unadulterated empirics and pure thought, lead man likewise to the fact that magnitudes are dependent on one another, that a change in one magnitude involves a change in another and oftentimes in many other magnitudes. This is the source of the movement which since the Middle Ages has been responsible for the function concept, and through its dominance has furthered its development. During the last decades it brought a change of mathematical education from a pure rigid number concept to a dynamic function concept. It is not necessary to elaborate further on this point because everything important in this respect was said by H. R. Hamley in the Ninth Yearbook of the National Council of Teachers of Mathematics.

Graphical representation. Luckily, in the teaching process the function concept is connected with its geometric representation as well as with its arithmetic illustrations. This relationship is of immense pedagogical importance; the number-feeling abilities of the pupils are strengthened and space perception is developed. The function concept is of scientific and mathematical importance also. Even Euler considered as a function that which might be presented in any manner by means of arithmetic symbols. Dirichlet first developed the function concept in the present general form, where any two variables are defined in a given relationship. We make direct use of this generalized function concept when, in teaching the graphical representation of an empirical function, we do not examine it but consider it from the methodological point of view.

Infinitesimal calculus. The development of the function concept still remains a difficult task if we do not introduce operations of the so-called infinitesimal methods. The introduction in schools of the differential and integral calculus has put in contrast as something mystical the development which had to do with the infinitely small and infinitely large. The pedagogy of infinitesimal calculus has now progressed so far that the fine definition of the infinitely small by means of finite averages as their limits, thanks to Cauchy and Weierstrass, offers little difficulty in the high schools. The importance of the results of the pedagogical work of the last decades cannot be overestimated.

Only recently has it become obvious how simple are certain things, such as the geometrical problems of the tangent or the area, and the concepts of velocity, acceleration, work, and force—to name only a few. Moreover, the infinitesimal calculus is an indispensable instrument in obtaining thorough knowledge, particularly in the study of definite organic growth.

Development of General Thinking Ability

The teaching of language. The proper use of language is the prerequisite for the formulation and transference of thought from one to another and as such it is necessary for the process of correct thinking. How a certain science states in words its achievements is of importance for itself, but it is also generally important for educational purposes and for practical and scientific thinking. Mathematical language is simple and direct. It avoids superficialities. It is essentially definite. Such directness would be equally commendable in other matters of daily life.

To the spoken language corresponds the written expression; clear, direct simplicity dispenses with the unnecessary but makes use of all that is required. What the requirements are for such a procedure in mathematical education are quite obvious and need no further elaboration.

The development of concepts. When it is the aim of mathematics and other exact sciences to go to the basis of things, then there should be the answers to the questions "On what is this based?" and "How is it further developed?" It has to do with definite concepts. We should then take great care that our concepts be definite in their applications. We should use logical definitions for concepts developed afterwards, and the fundamental concepts must be very carefully based on the foundation of their science. What we thus accomplish in general education is the clear realization of the necessity of concepts in everyday life: What is a meter, a kilogram?[6] How is our concept of time developed?[7] How do we determine the position of some place on the surface of the earth and what is meant, for example, by the net on a geographical map?[8]

Such questions lead then to the problems of epistemology. When one speaks of space, he may express it in terms of the eye or touch. Another may refer to the space of experimental or theoretical physics; again another, a disciple of Kant, may speak of space as pure perception and would believe it three-dimensional Euclidean. Then enters the mathematician and speaks of the non-Euclidean, affine, projective, or *n*-dimensional spaces. What does all this signify? When two persons concern themselves with the problem of space, the first prerequisite is that they clarify their concepts.

It would be fortunate if in mathematical education we could concern ourselves at least with the questions of foundations. This is out of the question in high school, but in colleges it should not be neglected.[9] Introduction to purer concept determination is already possible in school mathematics, in arithmetic as well as geometry, and at the very beginning—not by means of a logical deductive treatment, but as an inductive perceptual introduction.

Consecutiveness. The sequence of propositions should be cautiously examined in order to determine their logical relations. In mathematics the pupil learns by steps forward; by the connectives "therefore" and "thus" he arrives at the conclusion that one theorem is a consequence of another, and by "but it is" he is introduced to another fact. He learns to distinguish between a hypothesis and a conclusion. He learns the conditions for the formulation of a fact. He learns to distinguish between sufficient and necessary conditions. He learns the difference between hypotheses and statements. Briefly, he understands how to deal with situations in life even if they are not altogether simple.

[6] Pozdèna, R., "Mater und Kilogramm," *Math. Phys. Bibl. H.*, Teubner, Leipzig, 1934.

[7] Barneck, A. "Die Grundlagen unserer Zeitrechnung?" *Math. Phys. Bibl.*, 29, Teubner, Leipzig, 1932.

[8] Scheffers, G., "Wie findet und zeichnet man Gradnetze von Land und Sternkarten?" *Math. Phys. Bibl.*, 85-86, Teubner, Leipzig, 1934.

[9] See Lietzmann, W., *Grundlage und Aufbau der Mathematik*, Teubner, Leipzig, 1927; also Young, J. W., *Lectures on Fundamental Concepts of Algebra and Geometry*, The Macmillan Co., New York, 1925.

Responsibility. In the logical and correct relationship of propositions lies the proof. Other things being equal, the step from a simple conclusion with its three propositions to a more or less long chain of conclusions offers the pupil difficulties. The teacher should proceed cautiously and thoroughly. He should not underestimate at any rate the ability to absorb and to reproduce a proof.

Very important in this case is the awakening of the critical abilities, through emphasis on fallacies, incomplete proofs, and the like.[10] The pupil must be fully responsible for what he generalizes and for what he accepts on belief. In life we come in contact with two types of persons. One is very careful and reticent in his expression, but we may rely on what he says; the other—well, everybody knows him. Mathematical education is a very good means of upbringing, because in it there is no beating around the bush. Either what we generalize is true or it is false—there is no third case. If we cannot say whether it is true or false, then we must admit that fact. The teacher should in such cases fearlessly say to his pupil: "I do not know this and others do not now know."

Under no condition should anyone resort to authority in order to justify oneself. I like to address my pupils as follows: "You shouldn't say, 'It is said in the book,' " or "You shouldn't say, 'We learned that before.' "

Ability to combine. No mathematical education should avoid the step from a receptive reproduction of a proof to later and productive work, to the individual working out of originals. This may well occur in simple cases. The most important mental activity based on work thus done is the application of combinatory ability. We search various routes which lead to the same goal and test their practicability. Such procedure is often difficult because there is not merely one step, but a series of different steps. It is similar to chess in that not the single move but the possible series of moves that follow determines the good player.

I will not dwell on the famous transfer problem, but so much can be said: In practical life generally such combinatory ability in thinking is necessary, especially when there are no quick solutions of problems; for example, when long deliberation is necessary, when one must consider the reactions of others.

Scientific thought. Mathematics is an example of a science devel-

[10] See Lietzmann, W., "Trugschlüsse," *Math. Phys. Bibl.*, 53, Teubner, Leipzig, 1923; also "Wo Steckt der Fehler?" *Math. Phys. Bibl.*, 52, Teubner, Leipzig, 1923.

oped deductively. But whenever an exact science is pursued, it is first obtained inductively; singular cases are finally connected in a deductive system, even if at first we do it in partial fields, but mathematical thinking has a training value in general for exact scientific thinking.

For general education there is another value; that is, that we know the limitations of our thought. Nowhere is it so apparent as in mathematics that there are problems which have no solution, others which have not yet been solved.

Conclusion. Consideration of the educational value of mathematics gives rise to new problems: First, is such education possible? In other words, to what extent are mathematical abilities innate and to what extent can they be influenced by the surrounding world? Second, there is the problem of method. Third, there is the problem of the content of the mathematics curriculum. The question of the course of study should by all means never be considered from the materialistic point of view, as, for example, "What uses does the pupil make in his future life of mathematical knowledge?" As a rule this leads to a very poor solution. We should mostly determine the course of study in relation to the general educational objectives. But this opens up the question of the educational value of mathematics as the central point of the pedagogy of mathematics as a whole.

MATHEMATICS AS RELATED TO OTHER GREAT FIELDS OF KNOWLEDGE

By GEORG WOLFF
Dusseldorf, Germany

PART ONE
MATHEMATICS AND THE SCIENCES

Introduction

It cannot be denied that many people think of mathematics as an abstract, dull science concerned with points, straight lines, segments, curves, magnitudes, fractions, powers, roots, logarithms, and the like. They believe that mathematics is based on *rigorous and difficult proofs* (understood by only a very few) and also on certain fundamental ideas named *axioms*. Naturally one can trace back this view to the generations when Euclid and always Euclid was studied in schools. This study was done according to a dry, dull, and lifeless method, since discarded. The day of this method should have been forgotten long ago.

Much credit goes to the International Commission on Mathematical Education, whose work since about 1908 has influenced the methods of mathematical education throughout the entire world. We here in Germany, as well as Americans, even now vividly recall how David Eugene Smith, a historian of mathematics and a mathematics teacher by calling, collaborated with the International Commission.

Through such efforts a decidedly new course of study was introduced into the schools. We no longer teach the difficult proofs of the Greek mathematics in their abstract imperceptible forms; we search for the relation of mathematics to the life of the practical man and his workshop, the merchant who needs to perform many calculations, and the engineer who performs his calculations and drawings by means of technical tools. In short *applied, live mathematics* forms the central core of modern mathematical education. Instead of exact deduction we have many intuitively considered examples supplemented by numerous illustrations from many fields of applications.

Among these applications we should especially mention here those which appear in pure and applied natural sciences, in physics, chemistry, biology (botany and zoölogy),[1] as well as in the *arts*, such as music, painting, sculpture, architecture, and even poetry. We shall show the relation of mathematics to these sciences, but because of limited space we shall be able to show it in general outline only.

Mathematics and Natural Sciences

Natural philosophy. Not always have men applied mathematics to natural sciences. The school of Aristotle (384-322 B.C.) examined nature by means of the mind and therefore was satisfied with very few observations in order to arrive at a natural law as a generalization. This Aristotelian spirit dominated all natural sciences until the sixteenth century, when Galileo Galilei (1564-1642) formulated a new method. He showed that the inductive method of thought, such as that used by Aristotle himself, cannot lead to absolutely safe conclusions and cannot claim to be scientific; he required that observation and thought, experiment and number, induction and deduction, should be in agreement. His classical examples are *freely falling bodies* and *ballistic curves* in vacuum. Observations show that the distances through which bodies fall, S_1, S_2, S_3 . . . are in definite relationship to the time element; namely

$$S_1 : S_2 : S_3 : S_4 \ . \ . \ . = 1 : 4 : 9 : 16 \ . \ . \ .$$

This property Galileo generalized in the form of law only when it was expressed by means of the formula

$$S = \tfrac{1}{2}gt^2,$$

where g stands for gravitational attraction, and t denotes time. After this general property is discovered other conclusions may be arrived at; as, for example, the paths (w) traversed by a falling body in a single second which are in the following proportions:

$$w_1 : w_2 : w_3 : w_4 \ . \ . \ . = 1 : 3 : 5 : 9 \ . \ . \ .$$

Correspondingly we have the parabolic law $y = -cx^2$. This is geometrically represented by a parabola concave downward.

The novelty in this case was that Galileo formulated the founda-

[1] Hereafter the term "biology" shall be understood to include botany and zoölogy.

tions of mathematization of natural sciences, and this made him the founder of the exact natural sciences. Thereby he gave form to the method of natural research due to Plato (427-347 B.C.), and this became the common procedure in the succeeding centuries. It is still valid today. Mathematics is and should be in the full sense of the word the servant of the natural sciences. Thus number received a new meaning in natural research; the laws presented above represent functional relationship and dependence.

The most important problem of the new trend in the field of research in natural sciences is that the magnitudes which enter in a law should in reality be observed and obtained in agreement with the mathematical formula by means of which this law is derived. Whoever has performed experiments knows what it means to determine the time of oscillation of a pendulum, to evaluate the speed of the flow of gases, to observe the heat of a joule, or to measure the length of a light wave in terms of a 10^{-5} precision. The development of the experimental natural sciences has traversed the road of varied and improved scientific apparatus which has replaced the eye and the ear, and has enabled us to obtain measurements in close agreement with the formulas. *This road to precision, known to mathematics only, remains as a sign of progress up to the present day in all natural sciences.*

We must mention the fundamental investigations of observation and understanding performed by the rational thinkers throughout the whole world. The best way is to examine the history of philosophy as well as the various systems formulated by these philosophers. The important spiritual giants are found among the critical thinkers, such as Plato, Galileo, Descartes (1596-1650), Leibnitz (1646-1716), Newton (1643-1727), Kant (1724-1804); and among the devotees of the theory of knowledge of the modern times one may mention such men as Whitehead, Russell, Hilbert, Hermann Weyl, Planck, Heisenberg, as well as men who belong to various professional scientific circles.

In one respect, however, the help of mathematics to natural sciences has not been fully utilized. The natural law thus has a purely apodictical nature; it is 100 per cent certain. To this certainty *probability* was added during recent years. Thus, *not a single event but a multitude of events or complicated influences are considered.* Nowadays when we are dealing with the kinetical theory of gases, the laws of inheritance, the laws of attraction of masses, or the research in the constitution of atoms, it is universally accepted that a statistical interpretation is not only valid for one particular natural science but equally valid for all, whether it is physics, chemistry, or biology.

Physics. Naturally, we have no desire to disregard the methods of pure observation in natural research. These have their value, and have produced considerable results. We think of cathode rays —or X rays which were discovered solely by experiment; we think of automobiles and airplanes, of the use of motors, of the vast field of electrotechnics, of optical apparatus, and of appliances in the field of acoustics.

On the other hand, we have many phases of development in physics where mathematics and mathematical methods have produced strong impulses on research in natural science. One need only think of the steam engine. A new epoch began after a long standstill when the concept of *entropy* was introduced, when that part of energy which could not produce practical work was considered from the logical point of view.

Most results in physics are due to the development of *differential equations*, and especially to their modern advances, as well as to differential and integral calculus, the fundamentals of which should be studied in all high schools.

We are familiar with most of the many and notable advances of mathematical natural sciences. It is impossible to enumerate all of them.

So widespread was the skepticism concerning the expanding fields of the application of mathematics that multitudes were astounded when the Frenchman Leverrier, and the Englishman Adams, by observing certain perturbations in the movement of the planet Uranus in the sky, discovered mathematically a new planet; the Berlin astronomer Galle, in 1846, needed only to direct his excellent instrument in the place determined by calculations in order to announce the stirring news. It was approximately where Leverrier had predicted it by his computations, and it was the planet now known as Neptune. This glorious triumph of mathematical thinking cannot be repeated anywhere by anybody in the world.

We may add other illustrations. As the radio signals reach the smallest hamlets or the farthest corners of the earth, a physicist may recall the first step in the development of his science. The philosopher of natural science and physicist, James Clerk Maxwell (1831-1879), when examining the Faraday force-lines, discovered that electricity travels with the speed of light. This mathematical discovery was doubted by the scientists of his times; they considered Maxwell a confused phantastic man whose claims had no foundation. Experimental substantiation was lacking, and these ideas were considered impossible.

At that time the physical demonstrations which would have confirmed the prophesied electrical waves were not feasible. About a quarter of a century later, by risking his health and after many attempts, the well-known physicist Heinrich Rudolph Hertz (1857-1894) demonstrated these waves experimentally. But people were far from being able to harness these waves practically. It was more than twenty-five years before it was possible to put them to work. The differential equations of the light-waves and of the electrical waves stated in the same form had shown the theoretical physicist Maxwell the proper method. Thus, thanks to Maxwell, who gave us the electromagnetic theory of light, the unity of light-waves and electrical waves was demonstrated.

A brief examination of the latest phase in modern physics, the *quantum theory*, shows again the value of rigorous deduction. With the discoveries are associated the names of Max Planck and Niels Bohr.

Basically this theory had to do with the following: Concerning the rays of light of a candle, of an electrical lamp, or generally any ray of light, before 1900 two laws were advanced—one of Boltzmann, the other of W. Wien, both contradictory to the Maxwell electrodynamics.

Max Planck believed that the theory of Maxwell could not be doubted; therefore the fundamental concepts of transmission of energy (radiation) must be changed. He boldly proposed a hypothesis that energy consists of very small particles, similar to a table consisting of atoms or electrical waves of protons. This smallest element of energy was named an *energy-quantum* ϵ, or simply *quantum*. Planck then fearlessly assumed that this ϵ is proportional to the vibration number v of the respective way; so that we have the equation $\epsilon = bv$, where b is a proportionality coefficient (factor).

How fortunate this assumption was, one can judge from the meaning of the proportionality factor in *quantum theory*. It represents the *quantum* effect and its magnitude was experimentally determined. It is

$$b = 6.55 \cdot 10^{-27} \text{ erg. sec.}$$

By means of these assumptions Planck was able to reformulate the laws of radiation that were contradictory to Maxwell, and the two radiation laws agreed with each other. This discovery produced a tremendous effect. But it was still more important when Planck, starting in another direction, determined the number of molecules in a cubic centimeter of gas at 0° temperature and 760 mm. pressure as

$$N = 27.6 \cdot 10^{18},$$

which is known as the *Loschmidt number*, and which up to the present time is the best value known.

Besides this great discovery of the quantum theory there are many others, of which it suffices to mention four: the explanation of fluorescence, the explanation of strong and weak X rays, the study of the inconstancy of atomical heat, the Compton effect.

Of particular value to the discovery of the quantum hypothesis was the new atomical theory of Niels Bohr, who studied energy on the basis of Planck's assumptions, especially that energy radiated in any quantities but in integral multiples of ϵ.

The surprising conclusion gave clear ideas of the structure of atoms in the sense of the Keplerian planetary laws. We state them here:

1. *The paths of the electrons are ellipses. The central body is the atom-nucleus—it is located at the focus.*

2. *One electron may move about the nucleus along one determined path. This path can be computed by means of the quantum theory.*

3. *The electrons can jump from one path to another. When an electron jumps from one path to another nearer to the nucleus, then energy is released; should the path be further from the nucleus then energy is taken on.*

It only remained to study the paths of the electrons of atoms in the same manner as the satellites of the planets in the sky, but there is a distinct difference; namely, that the study in microcosmos is more difficult. It enabled us to formulate the inner structure of elements, and to discover a new element known as *hafnium.*

We have to mention also the new wave mechanics[2] as it was formulated by Heisenberg. The latest consequences of this quantum mechanics shook the strongholds of causality. And thereby statistics and probability entered into foundations of natural philosophy. The effect of a cause is only certain when the conditions of that cause are definitely known. If this is not the case, then we have to deal with *indeterminateness*, and the probability must be computed. There is no connecton between cause and effect from the point of view of certainty but only from the point of view of probability.

Thus we have reached the present limits of philosophy and of thought and of religion, concerning which the last word has not yet been spoken; for the quantum hypothesis gave rise to many contradictions which must be clarified.

After all, the quantum theory and the atomic theory represent a tremendous victory of the pure mathematical method in the field of physics. Many new problems were offered for solution to the practical physicist by Planck, Bohr, and their coworkers. We can judge the difficulties which confronted these workers from the following constants, which will subsequently be augmented by others.

1. The constant of the electrical elementary quantum: $1.6 \cdot 10^{-19}$ coulombs, or $4.77 \cdot 10^{-10}$ LE.
2. The diameter of an hydrogen atom (H): $12 \cdot 10^{-8}$ mm.
3. The weight of the H-atoms: $1.66 \cdot 10^{-24}$ gr.
4. The mass of an electron: $9 \cdot 10^{-18}$ gr.
5. The average separation between two collisions (H): $11.2 \cdot 10^{-6}$ cm.
6. The number of collisions (H) per second: $15.1 \cdot 10^{9}$.
7. The number of molecules in a cubic cm.: $6.06 \cdot 10^{23}$.
8. The weight of the earth: $6 \cdot 10^{24}$ kg.

Up to the present we have considered illustrations in which the mathematical investigation of the natural phenomena influenced the substance of subsequent experimental investigations. But the development of theoretical physics shows that experimental physics likewise influenced exact physics. The mathematical physicists had to base many assumptions on experimental physics. We know that in many cases the mathematician, in order to solve his problems, had to reorganize his method of analysis. Thus, in order to be able to study heat, Fourier developed his series; Einstein had

[2] Wolff, G., "Quantenmechanik und Kausalität." *Unterrichtsblätter für Mathematik und Naturwissenschaften*, 1931, 37, pp. 107 ff.

to master functions with an infinite number of variables; Laplace had to invent spherical functions so that he could solve certain astronomical problems.

The development of the *Law of Attraction* may serve as a very forceful example of the interdependence of experimental physics and mathematical physics.

Copernicus (1473-1543), with his fiery imagination, made a definite step toward the *heliocentric system*, according to which the sun was placed at the center of the world-system. Curiously enough, the astronomers in this new epoch, as in ancient times, had not yet formulated the behavior of the planets. Tycho Brahe (1546-1601) studied their motion by means of certain measurements, but he could not discover their paths, because according to Euclidean geometry these appeared to be circular, and as such they were accepted by all.

Johann Kepler (1571-1630), who studied Apollonius and his conic sections, recognized the elliptical paths, and found their substantiation in Tycho's tables. Thus he arrived inductively at his three laws.

But these fundamental laws of Kepler lacked rigorous foundation. The next important step was developed by Newton (1643-1727), who generalized the problem of the behavior of bodies throughout the universe. In this generalization he used behavior of two masses (m_1 and m_2). Thus he discovered his *Law of Attraction*

$$f = c \cdot \frac{m_1 m_2}{r^2},$$

where r is the distance between the masses, f is the force of attraction, and c is the famous gravitational or attraction constant.

With the help of this law Newton, by means of differential and integral calculus, proved from a purely mathematical viewpoint the correctness of Kepler's laws. Only after such proof did this law obtain full citizenship rights in nature, whereas previously its soundness had been doubted.

It remained for physicists to determine experimentally the gravitational constant. Experiments were repeated many times. The most recent value is given as

$$c = 6.66 \cdot 10^{-8}.$$

The meaning of Newton's law is universal. With it one can

explain the tides, explain the fundamental changes in the starry heavens, predict the motions of planets by correcting for the perturbations of their courses. Such was the state of affairs when Leverrier stumbled onto the discovery of a new planet by means of Newton's law. The theory of potentials in the gravitational field was evolved, and thus arose a new functional theoretical development in mathematics.

The culmination of current experimental development is found in the latest unification of the behavior of bodies (and thereby of Newton's law also) in Einstein's General Relativity Theory. The theory employs Newton's law as its first approximation. Its second approximation is represented by the thus far puzzling behavior of planets at the perihelion. Besides this, the theory holds that the rays of fixed stars, due to solar attraction, should be bent in the vicinity of the sun. Finally, the lines in the red part of the spectrum must be displaced in a gravitational field. These theories were substantiated by painstaking experiments.

So ends the historical excursion in the meaning of experimental physics, just where it was begun. Throughout, there has been constant interplay: induction and deduction, observational and exact physics, each contributing to the other, each forcing the other forward.

We cannot close this section without mentioning the unification of *space* and *time* by Einstein. Previously space and time had been endowed with properties of the *absolute*. In the unified natural science, relativity dragged in statistics. Strong and much too vigorous rationalism had to make concessions.

Chemistry. When one looks into a chemistry textbook these days, he immediately notices a great number of tables and graphical representations, none of which can be found in earlier texts. This change in textbooks reflects the development of chemistry. During the last two hundred years, at first rather slowly, then more rapidly, as a subsidiary science to medicine at first, it finally came to stand on its own, and thus from a descriptive science it emerged into an exact science.

This is not to be wondered at. The formula-language of chemistry is a kind of shorthand based on mathematical principles. Whatever the pupil has to do during the first few hours refers wholly and only to magnitude and numbers.

The total weight in a chemical experiment must remain constant.

The elements combine only according to definite weight relationships.

If there are several combinations possible between two elements, then their weight-relationships behave like integers. (The law of multiple proportion.)

The study of the laws concerning gases also led to formulas, which were gradually improved. These are:

Boyle (1627–1691): $pv =$ constant
Gay-Lussac (1778–1850): $v = v_0\ (1 + dt)$
Combined law: $pv = p_0 v_0 \cdot (1 + dt) = R \cdot T$
Van der Waals (1837–1923): $(p + \frac{a}{v^2}) \cdot (v - b) = c$

It is known that Avogadro (1776-1856) came across the law of Gay-Lussac. He discovered that it is possible to determine the molecular weight (M) by means of the pressure of steam (D). Thereby he introduced into chemistry further mathematico-physical methods. The molecular weight of any gas, relatively to the air, is

$$M = 28.94 \cdot D.$$

Avogadro arrived at it deductively. His hypothesis was: *The number of molecules of any gas in equal volumes is the same.* Thus arose the problem concerning the number of molecules in 1 cm.^{3} We mentioned above the Loschmidt number, which is also known as the Avogadro number; it is:

27.6 trillion molecules.

The next problem was concerned with the weight of the atom. Dulong (1785-1838) and Petit (1791-1820) expressed it by means of the law of atomic heat:

$$\text{(atomic weight)} \cdot \text{(specific heat)} = 6.4.$$

We may add here that the Faraday laws of electrolysis found their application to chemistry immediately after their discovery (1833). Thus we have the rise of quantitative physical analysis. We are now confronted with the epoch in which chemistry became an exact science. This was clearly shown in the investigations of the leading chemists. Faraday's teacher, Davy (1778-1829), was interested not only in chemistry but also in physics. He not only experimented in electrolysis but studied the electric arc between two coals. And Faraday, chemist by profession, became immortal through his discoveries in physics. His contribution made possible the development of present-day electrotechnics.

After chemistry took a definite course from a descriptive to an exact natural science, there remained the development of a definite scientific system. This tendency became obvious when attempts were made to discover generalizations of known laws which contained these known laws as special cases. And thus the mathematico-physical methods were more intensively applied.

Thus were developed chemical dynamics, the science of dissociation, the laws of the influence of mass, the structure of matter, and the periodic system of elements. The last is an abstract idea of Mendeleyeff (1834-1907). He based it on the group order of elements which were arranged according to their atomic weights. This system led in time to important discoveries. It made possible the prediction of the discovery of unknown elements. We mentioned the discovery of hafnium; but others were discovered in a similar manner: scandium, gallium, germanium, masurium, rhenium, and the numerous properties of these elements were predicted and finally experimentally observed. There cannot be a more wonderful example of deductive thinking than in this case of chemistry.

The history of disintegration of radium before Bohr introduced the atomic model is not only a part of physics but also a part of chemistry. The chemist must be familiar with the difficult mathematical theories which led Planck and others to the microstructure. Then only can he understand that:

1. *The chemical and physical properties of an element are at first determined by the electrons of the outer core. Elements having the same number of electrons have the same properties.*

2. *All atoms of a chemical element have similar* (*same*) *nuclei.*

Thus statistics and probability become mathematical tools of computation in chemistry.

But infinitesimal calculus thus becomes an essential tool for computation also. In the well-known book, Nernst-Schoenflies, *Einführing in die Mathematische Behandlung der Naturwissenschaften,*[3] appears the following statement:

[3] *Introduction in the Mathematical Treatment of Natural Sciences,* Oldenburg, 1923.

FIG.		MINERAL	CRYSTAL FORM
1.	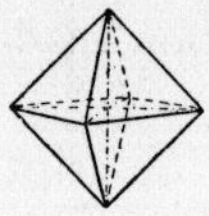	Magnetic iron	Octahedron
2.	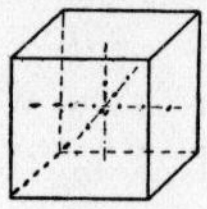	Salt, Lead	Cube
3.	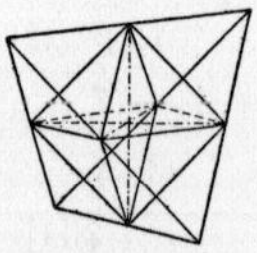	Zinc	Tetrahedron
4.	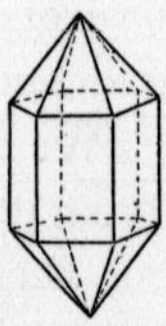	Quartz	Six-faced prism with pyramids placed on their bases
5.	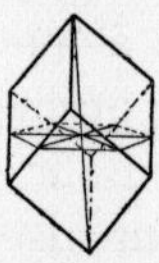	Lime	Rhombohedron
6.	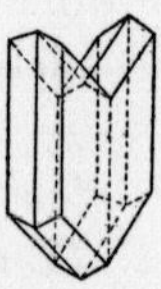	Gypsum	. . .
7.	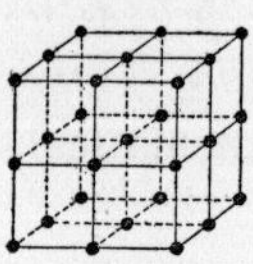	Metal atom	Cube
8.	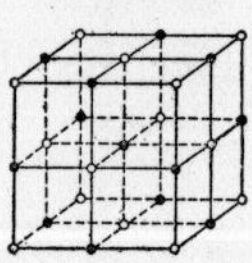	Salt atom	Cube

The chemist must constantly be aware of the fact that without the mastery of the elements of higher analysis theoretical chemistry will remain a book with seven seals. And if he wishes to avoid the danger of missing the opportunity of understanding all the developments in theoretical chemistry, a chemist should take care that the symbol of a differential and an integral cease to be mere hieroglyph to him.

Not only in number lies the strongest link between chemistry and mathematics, but in certain studies of form as well. The crystal forms of minerals are in most cases complicated mathematical bodies, as is shown in the variety of forms illustrated in Figs. 1 to 8 on the opposite page.

In conclusion we should mention the shell from geology (ammomite) with the mathematically constructed spiral, illustrated in Fig. 9 and Fig. 10 below.

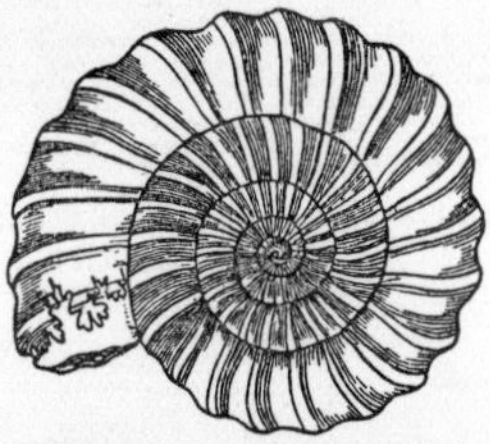

FIG. 9. Ammonite

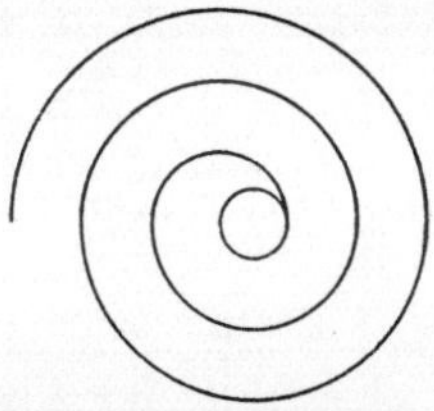

FIG. 10. Constructed Spiral

Biology. The biologists will never admit that mathematics has citizenship rights in their field. But biology would never have been a natural science had it not employed measuring devices and numerical methods, and had it not used them in order to express its laws quantitatively.

Apparatus. We therefore find many varieties of measuring apparatus, tables, or graphical representations in the field of biology. Fig. 11 shows one example, an apparatus which indicates the growth of a plant in a flower-pot. In Fig. 12 we have a graphical representation of the growth of a snowdrop plant in the period from December to March.

A very important idea in plant study is assimilation. We nowadays study the advance or retardation of assimilation by means of the different colors of the spectrum. We immerse the leaves of plants in alcohol. The extract is spectroanalyzed so that the absorption spectrum enables us to learn of the interchange of material. In Fig. 13 the analyses of both assimilation and absorption are plotted alongside each other so as to show their relation.

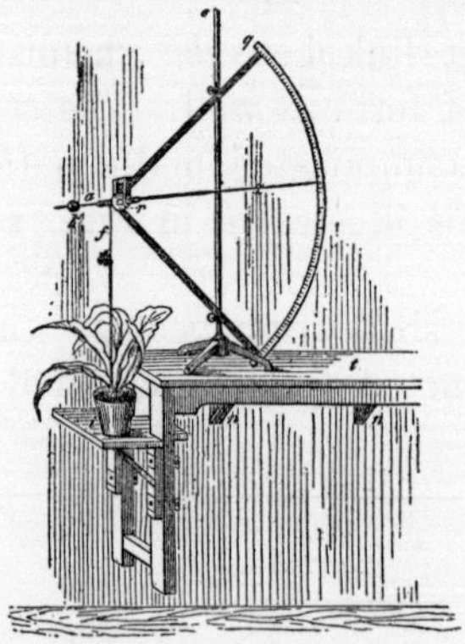

FIG. 11. Apparatus for Measuring Plant Growth

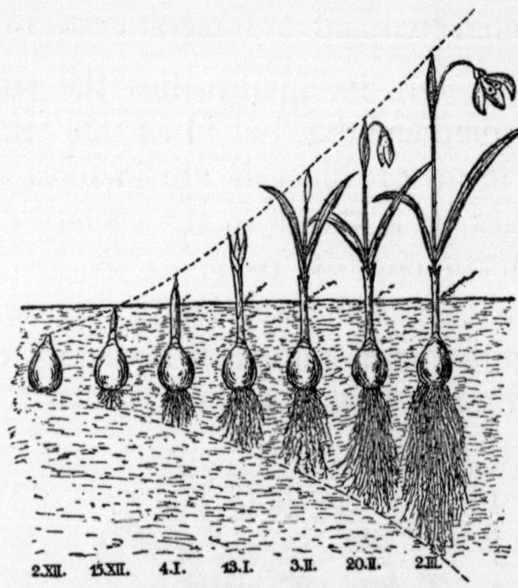

FIG. 12. Graphical Representation of Plant Growth

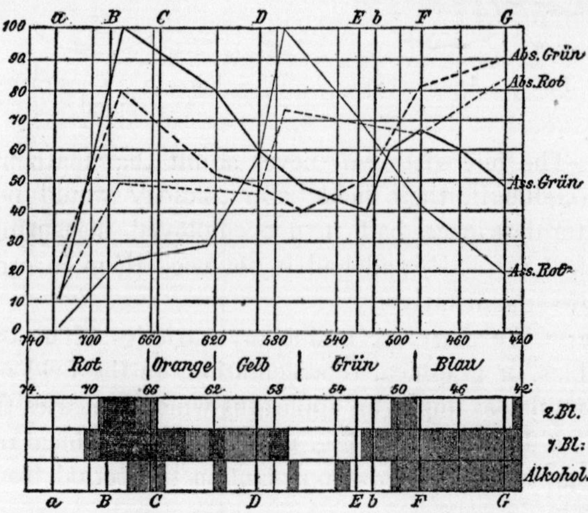

FIG. 13. Graphical Representation of Spectroanalysis of Assimilation and Absorption

Apparatus for the determination of the amount of absorbed carbon dioxide, as well as for the determination of the amount of acid material, is found in every biological laboratory. The graphic record reproduced in Fig. 14 shows how laboratory apparatus measures the effect of breathing exercises in speeding up the building of red corpuscles in man's body.

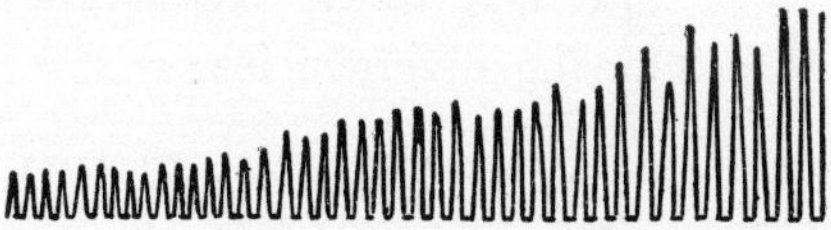

FIG. 14. Graphical Representation of Measurement of Metabolism

Measurements of temperatures of animals and plants, as well as functions of mass, surface, interchange of material, are performed by the finest apparatus.

The difficult measurement of pressure, such as that in capillary tubes in thin strands, or especially in roots, is thoroughly studied. The apparatus used in such measurements is illustrated in Fig. 15.

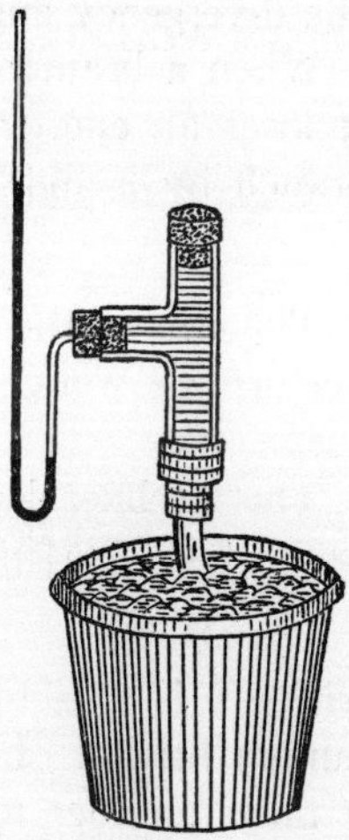

FIG. 15.

Finally, we should add mention of the study of the behavior of gases in cells of animals and plants. It follows an osmotic motion which is studied by means of physics.

Forms and Numbers. One of the most interesting fields of biology is the study of statics of cells and webs in the animal and plant world. That an ant in spite of its extremely small size is able to carry relatively heavy loads; that a very small bird may overcome with little fatigue a strong air current; that small, hollow bones of a small animal may withstand heavy strain of weight, that thin plants, such as grass, wheat, and the like, are constructed so strongly—these are matters that interest every friend of nature. And still we have not progressed very far in the study of the strength of the material of biologically small things; the resistance to bending, to twisting, and to pull are still to be studied thoroughly.

We describe the arrangement of leaves on plants by means of "divergencies." If in Fig. 16 the numbers 0, 1, 2, 3, the arrangement of the leaves succeeding one another, we find that 0 corresponds to 3. If we examine the divergency angle between the middle lines of two neighboring leaves, we find that it is 120°. This divergence is called 1/3, that is, we must go 1/3 of the stem further in order to reach the next leaf above. Generally we have the following divergencies:

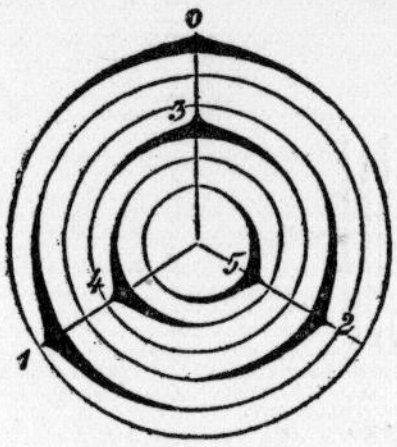

FIG. 16.

1/2, 1/3, 2/5, 3/8, 5/13, 8/21, 13/34, . . .

We observe the rule that the numerator and denominator of a new fraction are represented respectively by the sum of the numerators and the sum of the denominators of the two preceding fractions. The divergencies are between the limits ½ and 1/3, and the divergency angle approaches 137° 30′.

When a branch of a plant spreads out in smaller branches, then it loses an energy E. If we denote the energy of a new smaller branch by e and the angle between the two branches by α, then

$$E = e \cos \alpha$$

The size of the heart of a fish is proportional to its capacity for performing work. A trout, for example, has a larger heart than a carp.

The Fechner law gives the relation between sensation (E) and excitation (R) [in vision]. It is represented by the equation

$$E = k \log R,$$

where k is a factor of proportionality. This equation is valid for human beings. In the case of certain animals and certain plants it takes another form.

In order to show how biology leads into mathematics we may state the following problems:

1. The multiplication of animals, as well as humans, follows a geometric progression. In the case of a rabbit $a = 2$, $q = 7$; in the case of a house mouse $a = 2$, $r = 12$. What is the number of the animals of one generation after five years?
2. There are primitive animals which multiply by means of division into spores (for example, the germs of malaria). The division into spores takes place every two days. How many descendants of one cell are there after fourteen days, when in every division each spore divides itself into eight spores?
3. In the parthenogenetic multiplication of a plant louse only females of the species are born. During the summer there are on the average ten generations. What is the total number of individuals at the end of the last generation when each time the average

number of eggs laid is 300? After how many generations will there be 8,127,000,000 inhabitants?

4. A coccus [a germ] with a diameter of $1.2u$ divides itself under favorable conditions every twenty minutes. What is the number of individuals at the end of the sixth generation? How many cells are there after twenty-four hours? What is the weight in kilograms of a coccus-colony after twenty-four hours when the specific gravity of a live substance is 0.9?
5. The cylindrical bacterium of a gangrene has the dimensions: length $4u$ and width $2u$. The division takes place every forty minutes. How many bacteria are there developed after six hours? after two days?

A very important branch of biology is the science of heredity. We often desire to make permanent certain changes in plants and animals. This is only possible, however, through modification of many subjects. We investigate the size, the weight, the color, the number, the seeds, and other characteristics of many individuals of a species.

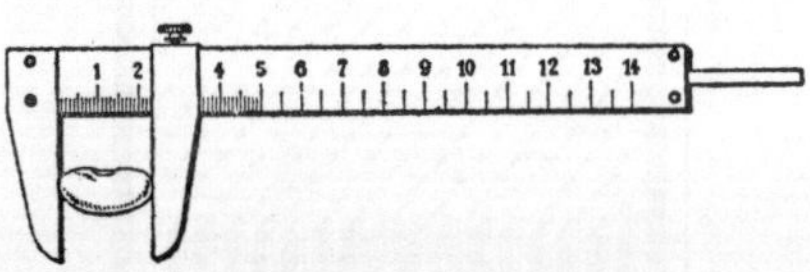

FIG. 17. Calipers

In this case statistical investigations only may be of help. If we wish to investigate the size of beans, for example, we measure their lengths with calipers (Fig. 17) and obtain a distribution like the following:

Length in mm.	10	11	12	13	14	15	16	17	18	19	20
Number of beans	1	7	15	26	60	70	64	32	13	8	2

If we plot these values on squared paper (Fig. 18), we obtain a variation curve (Gauss's error curve). The plotted curve deviates from the ideal curve, as in Fig. 19 for example, and it is necessary

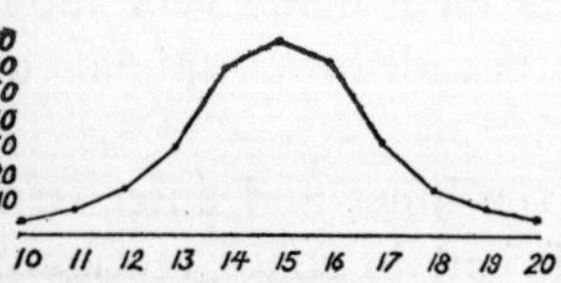

FIG. 18. Plotted Measurements

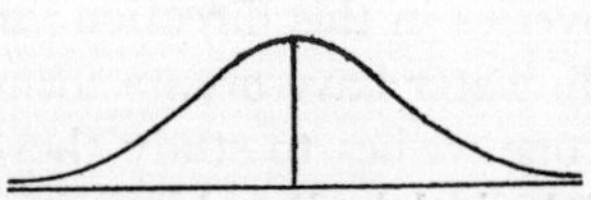

FIG. 19. Gauss's Error Curve

to determine this deviation. If Fig. 19 is carefully drawn, then it is the Gauss's error curve and is represented by the following formula:

$$y = \frac{1}{\sqrt{\pi}} \cdot e^{-x^2}$$

With the help of Galton's sieve (Fig. 20) it is possible to perform such investigations of collections of objects; the experiments yield the above-stated exponential function.

This method is found to be the best in all investigations of groups of men, animals, plants. Mortality tables of insurance companies are constructed along the same lines.

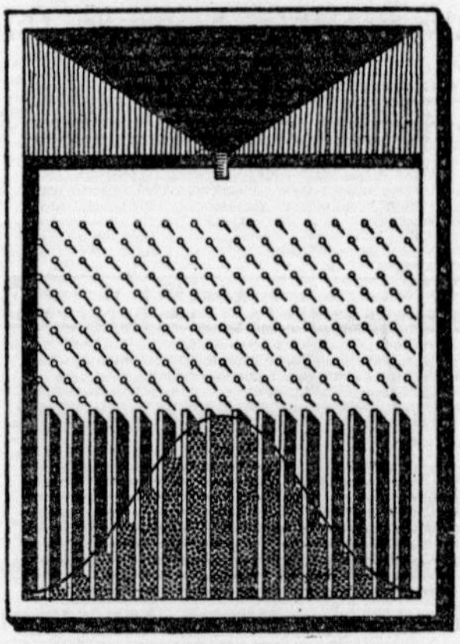

FIG. 20. Galton's Sieve

Form and Space. When we consider animal or plant forms, then we immediately notice the predominance of symmetry. Whether it is a leaf, or a flower, or a stamen, whether it is a cell, or a body of any animal, an artery, or an outer ornament, in many cases the left side is a mirror image of the right side. It would be an investigation in itself should anyone wish to study animals and plants along these lines. Haeckel and others have made very interesting discoveries in this direction; namely, that all lower animals on land and in water have ornate forms.

If one wishes to study thoroughly the inner and outer structures of biological bodies, he must be proficient in space representation. In this respect a knowledge of geometry and space perception is of great help to biological studies.

How expedient nature is in its workings can be seen from the example of a honeycomb (Fig. 21). Very careful investigations have shown that the bee is able to build bee-cells with a minimum amount of material; the cells are six-sided prisms with a superposed (at the bases) pyramid consisting of three rhomboidal plates. The natural scientist de Réaumur was the first to offer the suggestion concerning this maximum and minimum problem. Exact mathematical calculations by means of differential and integral calculus have substantiated the fact that the bee encloses a maximum of

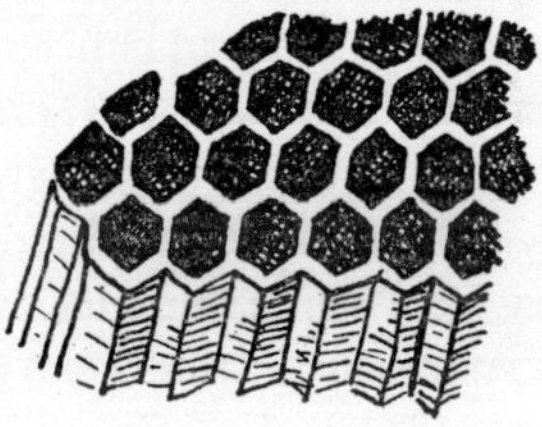

FIG. 21. Structure of Honeycomb

space with the minimum amount of wax. The rhomboidal plates meet at an angle of about 109° 28′.

This example of the bee-cell is especially significant for the assertion that mathematics should enter into biology. It is irrefutably certain, as the historical development of the connection between mathematics and physics and between mathematics and chemistry has so markedly shown, that from decade to decade, from century to century, the mathematical method of thinking, the mathematical method of number and space, the analytical method of higher mathematics, have increasingly become an invaluable aid in natural science. We mathematicians are proud of the fact that our science may help in the advancement of natural sciences. Many natural scientists may oppose this calculating tool and aid, but the day will come when the biologist will realize the amazing part played by mathematics in the progress of his science.

PART TWO

MATHEMATICS AND THE ARTS

Introduction

Since the nineteenth century mathematics has become an increasingly wonderful structure. Mathematicians[1] of all kinds, algebraists, geometricians, and those who investigate the theory of numbers and the theory of functions, repeatedly say in one form or another: *We, too, are artists.*

Whether one has a narrow or a broad conception of art, the mathematical method of working shows in its creative output a close relationship to that of the artist. Whether the one works with brush, with chisel, or with pen, whether the other works with numbers, with relationships between quantities, with configurations in a plane, in space, or in spaces, it is the creative imagination of each that is responsible for the finished product.

Fig. 22. Representation of an Elliptic Function

If one examines either analytically or geometrically certain mathematical creations—such as differential or integral equations, the theory of numbers, or Riemann's theory of functions, if one has before him the graphical representation of an elliptic function (Fig. 22) or a Steinerian surface or curves and surfaces in space (Figs. 23a and 23b), he will have to admit that research in these fields demands not only a keen intellect but also a real insight into and appreciation of the concepts of number and space. To the examples already cited could be added many others if the scope of this

[1] *Scripta Mathematica*, Vol. III, No. 2, pp. 103 ff; pp. 120 ff. Excellent articles by Nathan Altschiller-Court and by Cassius J. Keyser.

discussion permitted. Anyone who has a sense of rhythm or an aesthetic sense must admit that mathematics and art are related to each other.

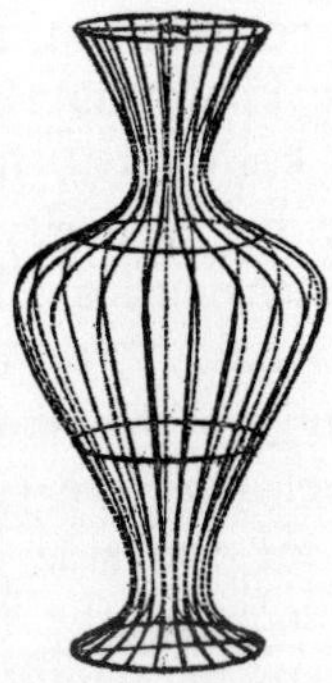

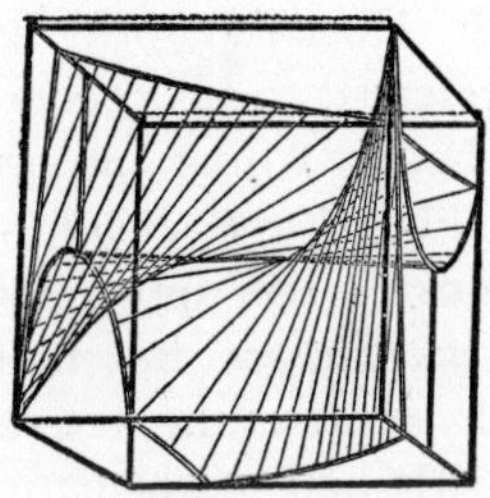

FIG. 23a. FIG. 23b.
Curves and Surfaces in Space

Artists have not been lacking in an interest in mathematics. We all know that the same person often has both musical talent and mathematical ability. The great Johann Sebastian Bach (1685-1750), his son Wilhelm Friedemann Bach (1710-1784), and the well-known Mendelssohn-Bartholdy (1809-1847) often occupied their leisure hours with algebraic and geometrical problems.

Of especial interest are the founders of the art of the Middle Ages. We know the famous group in Florence to which belonged the painter and architect Brunelleschi (1377-1446), the sculptor Donatello (1386-1466), the painter Uccello (1397-1475), and the versatile scholar Leon Battista Alberti (1404-1472). They were regarded not only as artists but also as scholars, and we know that they had a strong interest in mathematical problems. This fact is revealed by their published statements, and especially by their association with mathematical scholars, among whom were the mathematician and scientist Paolo Toscanelli (1397-1482) and the theologian and mathematician Nicolaus Cusanus (1401-1464).

We must still say a word about the artist and scientist Alberti. This year we are celebrating the five hundredth anniversary of the completion of the manuscript of his work, *De Pictura*—a work in which was attempted for the first time a scientific explanation of the principles of perspective, so important to the painter. He, who wished to serve art, busied himself with many mathematical problems, and with the principles of measurement and of surveying and the quadrature of the circle. Some of his technical problems of a mathematical nature have been published.

In certain ways Alberti reminds one of Leonardo Da Vinci. They were without doubt spiritually related—related, too, in their marvelous versatility and productivity. We know that Leonardo was familiar with Alberti's work, and we know that in Leonardo's writings he referred to his predecessor as his authority.

Leonardo had a strong interest in mathematics. Many of the drawings in his *Sketch Book* indicate that. For example, one sketch contains in the upper part a study of his most important work, *The Last Supper,* but in the lower part are two geometric figures which suggest the construction of the side of a regular octagon—artistic and mathematical problems on the same page!

Such harmonious propinquity is also to be found in the work of our great German master, Albrecht Dürer (1471-1528). There is a drawing by Dürer in the Dresden Library where a head is drawn at the bottom of the page, while at the top is the very difficult perspective drawing of a circle well worked out.

There are connecting links, too, between poetry and mathematics. Lessing, Goethe, and Schiller showed both in prose and in poetry an understanding of mathematical thinking; the romanticist Novalis sings of mathematics; literary men—Eyth, Hesse, Lasswitz, Morgenstern, Dominik, and others—discuss mathematical problems in their books.

Here we should mention again the new American publication already referred to, in which Cassius J. Keyser and Altschiller-Court have made important contributions. But above all is to be mentioned the outstanding essay of David Eugene Smith in his excellent little volume, *The Poetry of Mathematics and Other Essays.*[2]

We might close this introduction to the relationship between art (poetry, music, architecture, sculpture, painting) and mathematics with the words of Oswald Spengler in *Decline of the West.*[3]

> . . . The mathematics of beauty and the beauty of mathematics are henceforth inseparable. The unending space of tone and the all-round body of marble or bronze are *immediate* interpretations of the

[2] *Scripta Mathematica, Library,* Number One, 1934.

[3] *Untergang des Abendlandes,* translated by Charles Francis Atkinson. The bracketed number, following each excerpt specifies the page from which the quotation is taken.

extended. They belong to number-as-relation and to number-as-measure. In fresco and in oil painting, in the laws of proportion and those of perspective, the mathematical is only indicated, but the two final arts *are* mathematics, and on these peaks Apollinian art and Faustian art are seen entire. [284]

In the actual, tones are something extended, limited and numerable just as lines and colors are; harmony, melody, rhyme, and rhythm no less so than perspective, proportion, chiaroscuro, and outline. The distance separating two kinds of painting can be infinitely greater than that separating the painting and the music of a period. Considered in relation to a statue of Myron, the art of a Poussin landscape is the same as that of a contemporary chamber-cantata; that of Rembrandt as that of the organ works of Buxtehude, Pachelbel and Bach; that of Guardi as that of the Mozart opera—the *inner* form-language is so nearly identical that the difference between optical and acoustic means is negligible. [220]

. . . And so the born mathematician takes his place by the side of the great masters of the fugue, the chisel, and the brush; he and they alike strive, and must strive to actualize the grand order of all things by clothing it in symbol and so to communicate it to the plain fellow man who hears that order within himself but cannot effectively possess it; the domain of number, like the domains of tone, line, and color, becomes an image of the world-form. For this reason the word "creative" means more in the mathematical sphere than it does in the pure sciences—Newton, Gauss, and Riemann were artist-natures, and we know with what suddenness their great conceptions came upon them. [61]

Music and Mathematics

Oscillations and waves. Since we have asserted that music and mathematics are closely related, we will illuminate our statement with a single illustration. When J. J. Sylvester in the year 1864 said in his Trilogy: "Thus the musician feels mathematics, the mathematician thinks music," and when he writes of the spiritual relationships: Mozart-Dirichlet and Beethoven-Gauss, so we today see this relationship confirmed by pages of music where the artists have first written their compositions in curves before writing the music notes. It is certainly no accident that the "sine melody" or the "cycloid melody" may be discovered in so many compositions.

If one wishes to produce a tone, he does so through the vibrations of wires, little metal discs, and the like. These vibrations transmit themselves through waves. Each wave has a length λ, a time of oscillation T, and a speed of transmission C, (about 333 m/sec.). The relationship between these magnitudes is given by the equation

$$\lambda = CT$$

If one observes a wave, it is seen to resemble the sine curve. If t is the time of observation, then

$$y = a \sin\left(2\pi \cdot \frac{t}{T}\right)$$

is the analytical expression of the wave which is shown in Fig. 24.

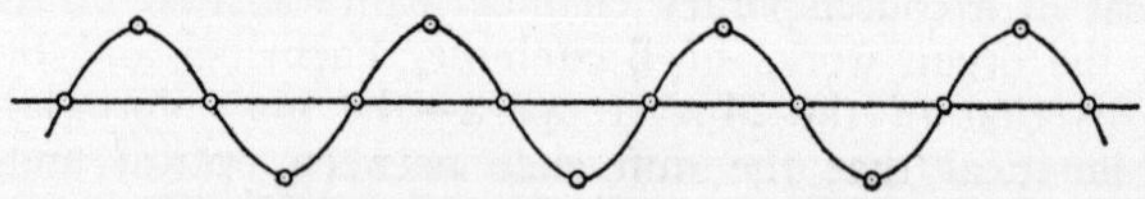

FIG. 24. The Sine-Curve Wave

By observing in a revolving mirror (Fig. 25) the vibrations which produce a tone, it is possible to get a graphical representation of

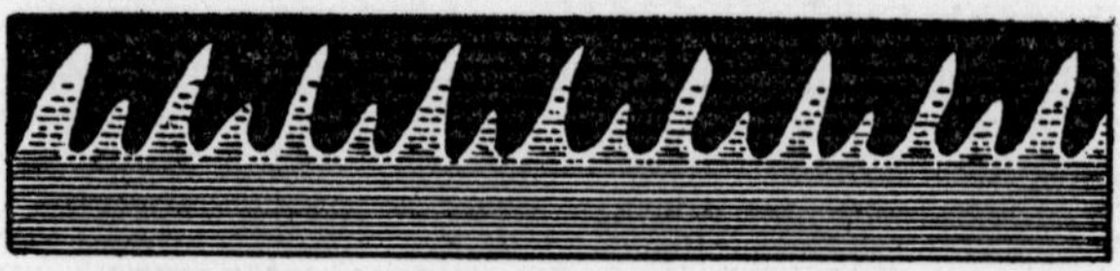

FIG. 25. Graphical Representation of Sound

the sound. Before we go to the scale, we must introduce the idea of *vibration number*

$$n = \frac{1}{T}.$$

That is,

$$c = n\lambda$$

The tonal scale. We start with the C major scale which has the notes C, D, E, F, G, A, H, C (Fig. 26). If one understands by *interval of two notes* the quotient of their vibration numbers, one sees that the vibration numbers of these eight notes are

$$24,\ 27,\ 30,\ 32,\ 36,\ 40,\ 45,\ 48$$

Consequently the intervals of the series are

$$\frac{9}{8} \quad \frac{10}{9} \quad \frac{16}{15} \quad \frac{9}{8} \quad \frac{10}{9} \quad \frac{9}{8} \quad \frac{16}{15}$$

We take now the ratios of each tone to the original tone (C), as follows:

1. $\frac{24}{24} = 1$ (keynote)
2. $\frac{27}{24} = \frac{9}{8}$ (second)
3. $\frac{30}{24} = \frac{5}{4}$ (third)
4. $\frac{32}{24} = \frac{4}{3}$ (fourth)
5. $\frac{36}{24} = \frac{3}{2}$ (fifth)
6. $\frac{40}{24} = \frac{5}{3}$ (sixth)
7. $\frac{45}{24} = \frac{15}{8}$ (seventh)
8. $\frac{48}{24} = 2$ (octave)

The diatonic scale has not always had the last-named interval. Pythagoras (*c.* 530 B.C.) who delighted in number speculation, placed them together as follows:

$$1 \quad \frac{9}{8} \quad \frac{81}{64} \quad \frac{4}{3} \quad \frac{3}{2} \quad \frac{27}{16} \quad \frac{243}{128}$$

or

$$1 \quad \frac{3^2}{2^3} \quad \frac{3^4}{2^6} \quad \frac{2^2}{3} \quad \frac{3}{2} \quad \frac{3^3}{2^4} \quad \frac{3^5}{2^7}$$

In the course of centuries this series underwent many changes, until on November 18, 1885 at the international music conference in Vienna, the Parisian concert pitch with the vibration number 435 (Fig. 27) was accepted as the international concert pitch. From it, with the help of the other intervals, the vibration numbers of all the notes of the C major scale and the preceding and following ones can be determined, and in addition the vibration numbers of the

FIG. 26. Piano Keyboard

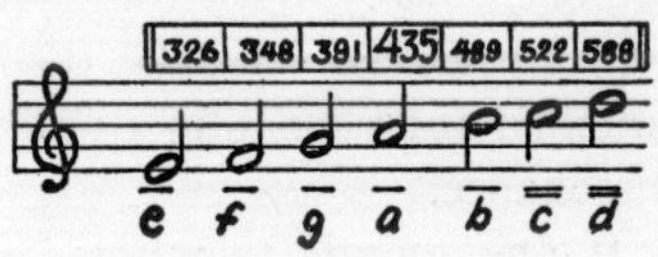

FIG. 27. Concert Pitch

notes advance from octave to octave according to the law of geometric series with ratio 2.

On the piano keyboard (Fig. 26) the white keys give the seven notes of the C major scale. But as the intervals between consecutive notes, as we have seen, are not equal $\left(\frac{9}{8}, \frac{10}{9}, \frac{16}{15}\right)$ the "black keys" *C* sharp, *D* sharp, *F* sharp, *G* sharp, and *A* sharp have been introduced between the great notes $\frac{9}{8}$ and $\frac{10}{9}$, now giving twelve notes to the octave with approximately equal intervals. Thus was developed what is known as the chromatic scale. In contrast to the chromatic scale, the seven notes of the scale have been named the diatonic scale.

With the use of the black keys, one comes to the minor scale *C*, *D*, *D* sharp, *F*, *G*, *G* sharp, *H*, *C* with their own tones, which unfortunately we cannot follow here. We will place the intervals together:

$$\text{Consecutive notes: } \frac{9}{8} \quad \frac{16}{15} \quad \frac{10}{9} \quad \frac{9}{8} \quad \frac{16}{15} \quad \frac{9}{8} \quad \frac{10}{9}$$

$$\text{Transposed to } C\text{: } 1 \quad \frac{9}{8} \quad \frac{6}{5} \quad \frac{4}{3} \quad \frac{3}{2} \quad \frac{8}{5} \quad \frac{9}{5} \quad 2$$

If one examines the intervals of the chromatic scale, he finds, as we have already said, that they are approximately, but not absolutely, equal. On the basis of this number investigation, musicians have adjusted these intervals. The necessary correction has been called temperament. The method by which musicians made the twelve notes equal was named equally balanced temperament, or equal temperament. One speaks, then, of the tempering of the scale, and of the tempering of the piano. Every interval then amounts to

$$i = \sqrt[12]{2} = 1.0595\ldots$$

Harmony and melody. Until the early Middle Ages music was for one voice. With the development of church music part-singing was introduced. In the theory of harmony, the euphony of notes sounded together was studied, and one speaks of consonance and dissonance.

If one sounds two notes together, the simpler the ratio of the numbers for that interval the more harmonious the sound. For example, if one strikes together consecutive notes of the *C* major scale, the sound will not be as agreeable as that of a note combined with the third or the fourth or the fifth note following. This observation leads us to follow up the intervals in detail. For pleasing sound of notes together, the triads given on the opposite page must be observed.

Major triad or major chord or major tonic chord, $1 : \frac{5}{4} : \frac{3}{2} = 4 : 5 : 6$

Minor triad or minor chord or minor tonic chord, $1 : \frac{6}{5} : \frac{3}{2} = 10 : 12 : 15$

Major-sixth-tonic chord, $1 : \frac{4}{3} : \frac{5}{3} : = 3 : 4 : 5$

Minor-sixth-tonic-chord, $1 : \frac{5}{4} : \frac{5}{3} = 12 : 15 : 20$

The figuration of notes gives to a piece of music the melody in the widest sense of the word. Parts of the composition yet to be considered are the rhythm, the time, the rest, the measure, the end or tonic accent, and the end or end tonic chord. These, too, can be and are expressed in numbers.

These suggestions do not exhaust the numerical possibilities of the notes and their combinations. One could still give the intervals of the different major and minor scales; one could speak of tones and overtones; one could aim high and express the relationships in formulas. Then, too, one could go into the exact calculation of the vibration numbers. But we must put all of that aside in order to consider number relationships in other graphic arts.

Proportion in the Graphic Arts

Architecture. In the last section we referred to Pythagoras as the man who had studied number relationships in music. We must now refer to him again in connection with the graphic arts; for from him came the statement: *Art is number.* Modern scholars have said: *Architecture is frozen music.* More significant is the statement of the Duke of Urbino, Federigo von Montefeltro, because it reveals an appreciation of geometry in architecture:

> Architecture is founded on arithmetic and geometry, which belong to the foremost of the seven free arts because they carry within themselves the highest degree of certainty.

We know that the circle was held in high regard by the ancients. For a long time it dominated geometry; it dominated the Romanesque style which flourished in the tenth and eleventh centuries. Circular windows, circular arches, even circular cross-vaulting predominated. Fig. 28 shows the importance of the circle in the window by Villard de Honnecourt, French artist of the thirteenth century.

The Gothic style broke away from this and created the pointed arch, illustrated in Fig. 29. In the Gothic arch is to be seen the

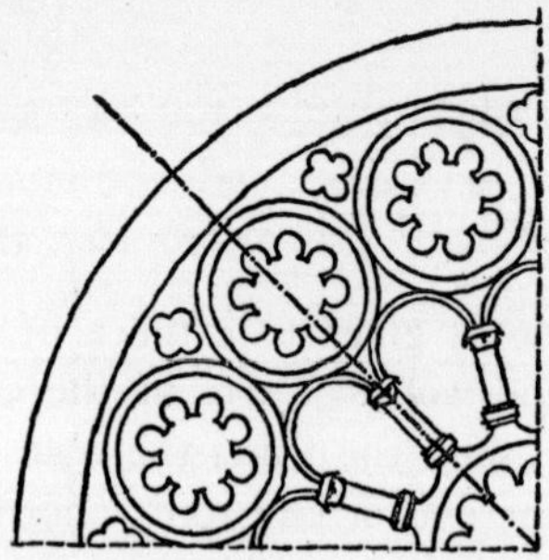

Fig. 28.

transition from the circular to the curve. Its supreme expression is found in the dome of the Cathedral of Florence built by the Florentine architect, Brunelleschi (1377-1446). Brunelleschi renounced the

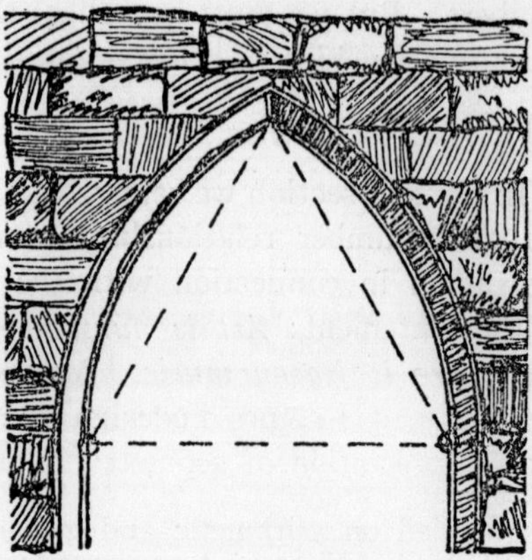

Fig. 29. Gothic Arch

pointed tower of the Romanesque and Gothic styles and created the dome. We know that the architectural construction, the calculations, caused him much mental strain. We know that conversations with his mathematical friends turned him in the years 1420-1434 to actual construction of the dome. The outer form of the dome was also an innovation. For the first time a catenary (Fig. 30) was used as a cross section of the dome of a cathedral. The equation of a catenary is:

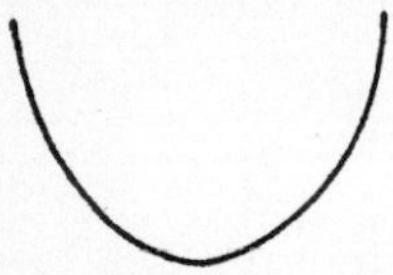

FIG. 30.

$$y = \frac{a}{2}\left(e^{\frac{x}{a}} + e^{-\frac{x}{a}}\right)$$

which must be seen as a transition to the parabola. Galileo himself at first thought the catenary was a parabola. If one rotates the inverted catenary around its axis of symmetry, he has the contour of the dome. Fig. 31 shows the constructed dome. The dome stands as a symbol at

FIG. 31. Cathedral of Florence

the beginning of the Renaissance. The curve and the straight line were its elements, but above all the line.

Common to the Romanesque, the Gothic, and the Renaissance styles is symmetry. The Roman architect Vitruvius has handed down to posterity in his ten books about architecture the tenet that not only must every building possess right and left symmetry, but the entire structure must also be ruled by this harmony. The mathematician rejoices when the architect or the art historian speaks of the functional tie-up of the entire structure with its parts, if rhythm is to permeate the building. What the temples of Rome began we see throughout the Middle Ages in Italy, in Germany, and

FIG. 32.

FIG. 33. The Louvre

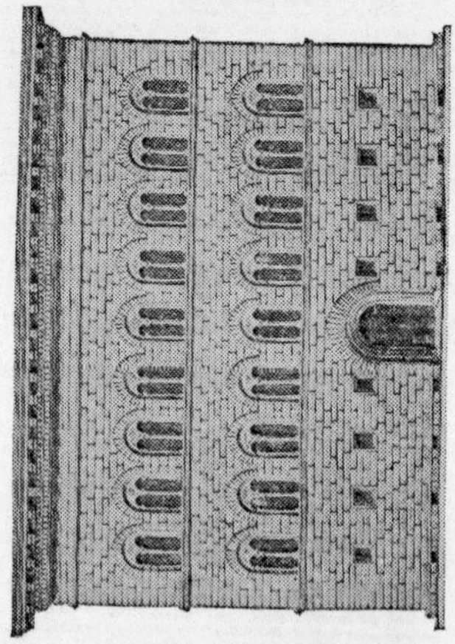

FIG. 34. Palazzo Strozzi in Florence

in France with fixed regulations for the ground plan and façades of churches, palaces, and residences.

But mathematics goes still deeper. The great masters of Italy and of other countries gave their pupils exact instructions for the drawing and the arithmetical construction. During the Middle Ages in Germany the *Bauhütten* (foreman's offices in a building yard) played a great rôle. Their directions were regarded as secrets (*Hüttengeheimnisse*). Only a few details could be mentioned. Houses several stories high were, in general, so built that the height of each story was greater than the one just above it because it had a greater weight to carry. However, there are many buildings where only the first story is higher than those above it, the rest being of equal height (Fig. 32). In the French Baroque

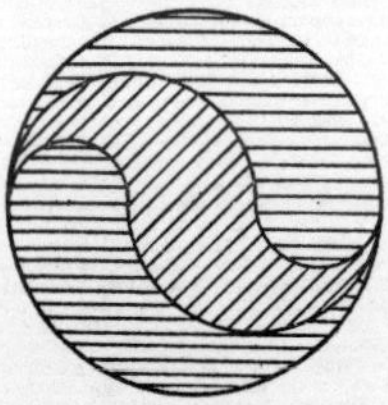

FIG. 35. Fischblasen

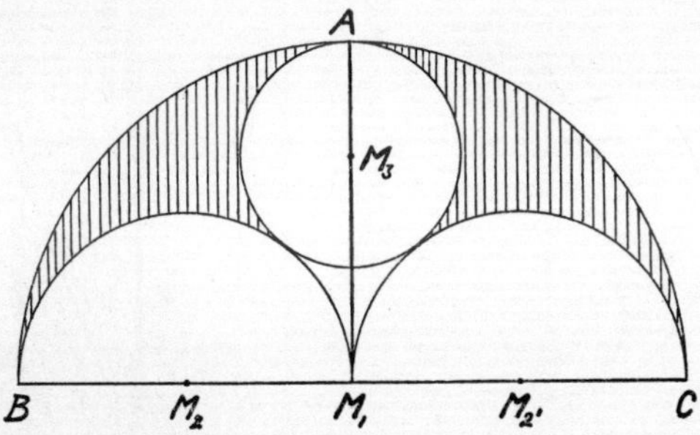

FIG. 36. Romanesque Window

style the upper stories markedly diminish in height. As an illustration, we will examine a part of the west wing of the Louvre (Fig. 33). The front stories are arranged almost like the terms of a decreasing arithmetical series. In the Renaissance the heights of the stories of the front elevation do not diminish so rapidly. The great architect-theorist, Sebastian Serlio (1475-1552), has told us that according to his opinion the terracing of the front stories from the first floor up should follow the law of the geometrical series where $r = \frac{3}{4}$, while with the famous Palazzo Strozzi in Florence (Fig. 34) the ratio is $r = \frac{5}{6}$.

The construction of church windows has always been a difficult problem. *"Fischblasen"* [in the form of a fish bladder] (Fig. 35) and circular windows (Fig. 36) were favored. It is a stimulating exercise for a mathematics class to compute the radius AM_3 from BM_1. The result is:

$$AM_3 = \frac{1}{3} BM_1$$

In the Gothic style there is an abundance of tracery in the decoration of the windows. Especially characteristic is the round trefoil (Figs. 37 and 38). The construction of this ornament is based upon the equilateral triangle.

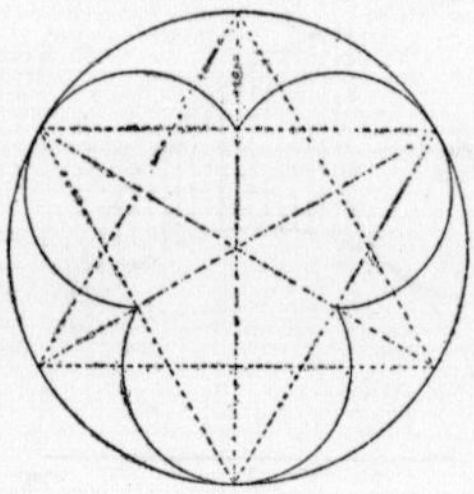

FIG. 37. Round Trefoil Construction

FIG. 38. Round Trefoil Window

The equilateral triangle plays an important part in Gothic architecture. We have already met it in Fig. 29 with the pointed arch. The masters used it in many places—from the ground floor to the turrets. There had also developed some knowledge about the application of the so-called $\frac{\pi}{3}$ triangle, the rigorous proof of which is still to be undertaken. We shall not attempt to discuss it here; for it appears partly mysterious and speculative.

We are quite certain, however, about the construction of the pillars. The *Säulen-Ordnungen* (order of pillars) gives the mathematical construction to the smallest detail. We must at least mention the Säulenbuch of Vignola (1507-1573), which gives the ratio of the height (h) to the diameter (d) of the lower base:

Doric columns:	$h = 8d$
Ionic columns:	$h = 9d$
Corinthian columns:	$h = 10d$
Tuscan columns:	$h = 7d$

While the artists of that period concerned themselves with proportion and form in architecture, Raphael (1483-1520) bemoaned that Vitruvius had thrown too little light on the architecture of the ancients.

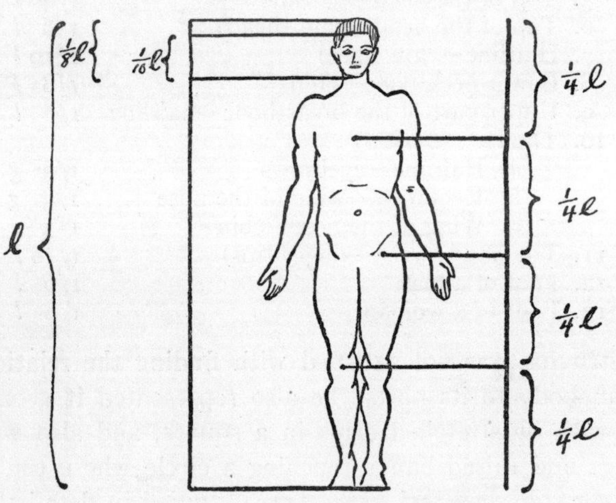

FIG. 39. The Canon of Vitruvius

Sculpture and painting. Michelangelo (1475-1564) recommended to the architects of his day that they master the necessary tenets of proportion through a study of proportion in the human body. One sees from this how prevalent at that time was the idea of "man as the measure of all things." The oldest source which we have concerning the proportion of man is *De Architectura*, the previously mentioned ten-volume work of the Roman architect, Vitruvius. In the third volume Vitruvius compares the temple building with the figure of man: the building which serves for the glorification of the gods must have as flawless proportions as the most perfect thing created by the gods themselves.

Thus it came about that the measurements of man were recorded in a table called the *Canon*.[4] It contains the proportional numbers for all parts of the body. Over this Canon, however, hovered as the first law that of symmetry as the harmonious foundation of that concept of beauty of the ancients and of the Middle Ages.

Vitruvius based his fractions on the total height, l (Lange) of man, the face, g (Gesicht), and the head, k (Kopf) (Fig. 39).

THE CANON OF VITRUVIUS

1. Top of the head—chest	1/4 l
2. Chest—upper part of the thigh	1/4 l
3. Upper part of the thigh—knee	1/4 l
4. Knee—sole of the foot	1/4 l
5. Top of the head—hairline	1/40 l
6. Top of the head—chin (head)	1/8 l
7. Hairline—chin (face)	1/10 l
8. Lower neck bone—chin	1/8 l
9. Upper part of the breastbone—hairline	1/8 l
10. The face (*Gesicht*)	
(*a*) Hairline—eyebrows	1/3 g
(*b*) Eyebrows—wing of the nose	1/3 g
(*c*) Wing of the nose—chin	1/3 g
11. The hand (wrist—finger tips)	1/10 l
12. Foot of a man	1/6 l
13. Foot of a woman	1/8 l

But Vitruvius was not satisfied with finding the relationship of the human body to its parts; he also represented it geometrically standing with outstretched arms in a square, and also with wide-spread legs and raised hands touching a circle whose center is the navel. Leonardo da Vinci (1452-1519) studied in detail the Canon of Vitruvius, and from the study he drew a sketch in his book which Vitruvius previously had drawn as two figures.

The Gothic style brought from Villard de Honnecourt the geometrical construction of the face (Fig. 40a) and of the body (Fig. 40b) based on the square, the rectangle, the equilateral triangle, and the isosceles triangle.

In this historical chain belong, too, the measurement tables of

[4] The Canon is the table used in art, painting, or designing. In the ratio employed, the denominator is the length of the human body, the numerator is the length of a part of the body. For example, if the distance from the top of the head to the chest is 50 cm., the length of the body 200 cm., then we have the proportion $\frac{50}{200} = \frac{1}{4}l$, where l is the length of the body.

Cennino Cennini (*c.* 1400), of Lorenzo Ghiberti (1378-1455), of Filarete (1400-1469), of the mathematician Luca Picioli (1445-1514), of the sculptor Pomponius Gauricus (*c.* 1504), to name only the most important.

The course of development shows plainly the one line from the still unmentioned Egyptians—who drew their people and animals on squared divisions—to Vitruvius, Villard, and many others on into the fifteenth century. All have made their canons in order to give to their gilds, to their pupils, this instruction: Thus and only thus may you proceed with the graphical representation of man.

FIG. 40a.

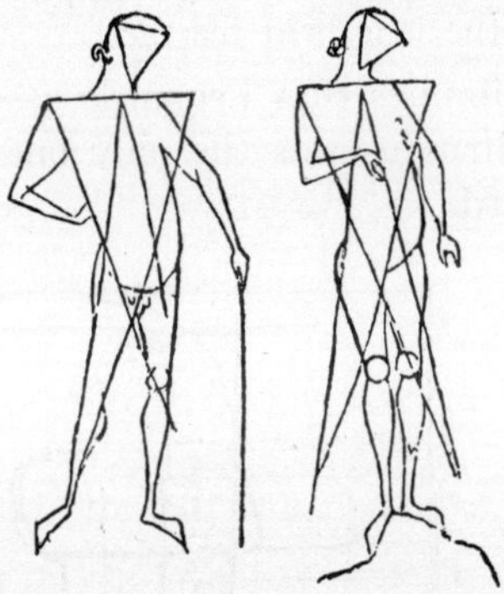

FIG. 40b.

Construction of Face and of Body, by Villard de Honnecourt

The course changes when Leon Battista Alberti (1404-1472) whom we have previously mentioned, wrote his *Anthropometrie: De statua* which must have been finished in the years 1464-1472. With this art-theorist and versatile scientist a new epoch in art began. His goal is that of the Renaissance: to unite art with science, to free it from the grasp of technicalities, and to strive for an ideal standard of beauty. He gave, therefore, no ratios, but absolute numbers; he introduced his own rule and a measuring instrument with which the most exact measurements could be made. His many measurements he kept in his canon. He continued to strive for greater exactness, for objectivity in the sciences! But it was contrary to his conception that his number series should be blindly accepted; he demanded of other artists that they make their own measurements, that they not do mechanical painting but decorative painting, that they not employ the art of a book of rules, but that of the artistic eye. So he strove toward a warm, picturesque style in art, toward beauty!

Leonardo industriously studied Alberti's views and digested them. If in his canon he also followed Vitruvius, it was in an artistic way and far from mere mechanical construction. The fact that he did not think of a table of measurements for adults only showed that he wanted to see with an artistic eye, that he wanted to develop an ideal of beauty. In his treatise he said:

> In early childhood the width of the shoulders is equal to the length of the face . . . but when man has attained his growth, each of these has developed. . . . The height of a well-proportioned adult is ten times the length of his face.

That implies that Leonardo assented to the divisions of Vitruvius; for Vitruvius was the only one who used the ten divisions before Leonardo.

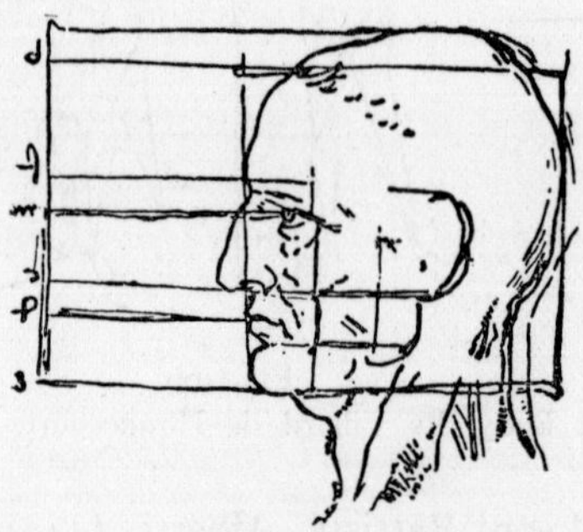

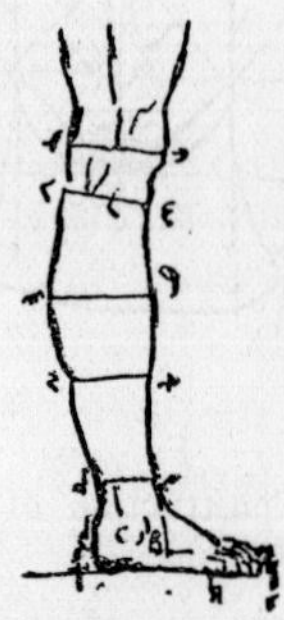

FIG. 41a. FIG. 41b.

Construction of Face and of Leg by da Vinci

It was not Leonardo's way to work a long time on one problem. His head was too full of ideas for him to be able to work at each problem systematically. So he leaves us no comprehensive summary; one must collect the numbers piecemeal. He likewise left no table of numbers concerning proportion in the human form. We have several sketches like those of Figs. 41a and 41b. We recognize plainly that the face is trisected by the brow, the nose, and the mouth and chin. We see that other parts, as the eye and the upper lip, are in proportion. The same is true of the drawing showing the shank.

In Germany Albrecht Dürer (1470-1528) gave close attention to proportion. He also wrote an extensive treatise about it. It would be interesting if we could examine his conception here in detail, but we will seek to clarify his goal by discussing several of his pictures.

The numerous lines in one of his figures show how the artist has striven to divide the body into a still greater number of parts in order to achieve a still more accurately constructed figure. The exactness was Dürer's own, inasmuch as he was an artist of the Renaissance period.

Dürer did not restrict himself to the arithmetical divisions. He repeatedly made geometrical constructions. He tried to represent the human body not only at rest but also in motion.

How rhythmically he proceeded with his composition is shown in a couple of drawings of the human hand and face. In the figure of the hand one can see the lines that Dürer himself drew. In the drawing of the face the lines were copied by Herr Justi.

Dürer also made a scale drawing of a horse according to his own method and according to Leonardo's in order to be able to recognize how, even in the animal kingdom, a standard of beauty was striven for through the canon.

Perspective. There are two problems with which the artists of the Renaissance occupied themselves—the problem of beauty and the problem of three-dimensional representation. The first was solved through proportion, the second through a knowledge of perspective. Brunelleschi, Alberti, Donatello, Uccello, Leonardo, and Dürer shared in this development.

To understand perspective, one must compare the picture which one sees with the object itself. For example, if we look down a railroad track, we see the tracks, telegraph poles, trees, and perhaps a river. In reality the rails are parallel, the telegraph poles of equal height, the trees of equal size, and the river of practically constant width. But in a picture of this view the rails apparently run together in a point which we call the vanishing point—call it H; the telegraph poles become smaller and smaller in the distance and also converge toward H; the river appears in the form of a triangle with the vertex in H; the trees on the mountainside appear to have diminished in size at H. Such a picture is called a perspective drawing of the object. All of the receding lines meet in one point, H.

If a painter wishes to depict space, he is expected to choose a central point. Painters have not always done that. Before 1400 it was not done at all; and even after the Florentines had found the point H, it was a long time before it was universally well done. In the works of Konrad Witz (*c.* 1447), a well-known painter of Basel, the receding lines do not converge. The painter did not construct his framework, but created the picture as he felt it to be. For that we cannot reproach him, for artistic freedom is due him; and we know that impressionism, expressionism, and cubism have created some fine works of art without the artist's submitting to the laws of perspective.

Quite different was the famous teacher of Raphael, the painter Perugino (1446-1523). In his *Annunciation* we see all the receding lines in the lower picture rapidly converging at a point. The same is true of the two upper pictures at the sides, as one may observe by drawing the lines.

In the year 1514 Albrecht Dürer completed the famous engraving *St. Jerome in His Study*. This time H (the point of convergence) is not in the center of the picture, but to the right. Without central perspective and without symmetry the artist has succeeded in creating an impression of great inward calm and of spiritual accord.

One frequently finds the opinion expressed that it was only in the Renaissance period that a central lineament was employed as the foundation of space. This is not the case; in every century, even today, there are many artists of reputation and eminence who have deviated from such practice.

Peter Janssen (*c.* 1682) had always observed the vanishing point in his pictures; in his *Interior of a Dutch Home*, as for example, the floor lines and the wall lines all are directed toward a point (H).

Observance of the central point is more difficult in sculpture. Donatello created exquisite architecture-plastics or art-sculptures. Thus in some of his works one can see the receding lines wonderfully centered, as they must be in relief perspective.

From such analyses and reconstruction of drawings, one can acquire an artistic point of view. It offers to mathematics a clarifying approach to art education and to art itself.

FORM AND APPRECIATION*

By GRIFFITH C. EVANS
University of California

It was Newton's modest disclaimer that he was not a great man, but stood on the shoulders of great men. We question rightly the first clause of this remark, but the last part seems as certainly true as the first part is false; for, like acrobats in the realm of time, we see the mathematicians of the present day standing on the shoulders of those of the previous generation, and so on back, for many centuries. Newton does not climb with difficulty to the height of Galileo, but is almost born there; and Newton's successors have the same advantages over him as he over his predecessors. It is this characteristic of mathematics, that what has been done in the past remains valid and important in the present, which seems to set it towering above other branches of knowledge. Moreover, mathematics does not end with Poincaré and Einstein.

The modern painter regrets that he could not have been born during the Renaissance and taken part in the vital discoveries of technique that were made at that time—of perspective, of the treatment of light and shade, of accurate anatomy, and so forth. In mathematics, that lament is inconsequential, for mathematicians in each age stand at the level of new things, and it is for each person an individual matter what new height can be reached. There is room on the shoulders of great men for us all to stand and peer out into the unknown.

This may seem to be an undue exaltation. Some tell us that mathematics is that science by means of which anything may be proved, whether or not it chances to be so, or they tell us it is that something, *not* science, in which one does not know what he is talking about, or whether it is true or not; and they suggest perhaps further that nobody else cares very much. Mathematicians indulge in endless combinations of symbols, develop hair-splitting

* Address delivered at the annual meeting of California Chapter Alpha of Phi Beta Kappa, March 7, 1935, at Berkeley.

distinctions about the properties of lines that have no breadth and points that have no size at all, imagine figures in a fourth dimension, and spend their time in various ways on things that do not exist. What difference does it all make in a busy world?

Yet evidently this is a very incomplete account of mathematics. For our whole lives, from paying a gas bill to designing a piece of machinery, are dominated by it. The revolution of the planets and the flux of the tides are subservient to it, the molecules and electrons are under its sway. And also, perhaps for that reason, it provides an opening through which the mathematician may look at the world. I wish to explain what I see, through this high window, of some of the things of the world of spiritual values, and what mathematics seems to teach me of the creative activity of others.

We are all familiar with the fact that there is such a thing as non-Euclidean geometry, and we know that it has something to do with the unsuccessful effort, lasting over several centuries, to prove the ancient and time-honored statement that through a given point, not on a given line, there could be drawn one and only one parallel to the line. As it turned out, this effort was unsuccessful for a very good reason; namely, that, as was finally discovered, systems of geometry, just as sound as the customary one, could be built up in such a way that the statement would not be true. The geometry which resulted was about as real as before, in the sense that it could be interpreted in some sort of fashion in space. In fact, one of the geometries of this kind could be applied to a portion of the surface of a sphere, and its theorems were familiar as solutions to the problems of geodetic surveying. But in this way, the emphasis in geometry came to lie in its nature as a combination of axioms and deductions, rather than in a study of actual figures in space. Thus a geometry of four dimensions, or of any convenient number of dimensions, could also be developed systematically out of axioms and definitions.

From this kind of development, however, there arose an entirely new attitude towards mathematics. It is this attitude which I wish to explain briefly, and then, to some extent, to combat.

In projective geometry of space the axioms are so stated that there is a duality: if the words "point" and "plane" in any axiom are interchanged, the axioms as a whole are the same as before. For example, if we write the two axioms (1) "two points determine a line" and (2) "two planes determine a line," the change mentioned obviously does nothing but interchange the two axioms, and the system is the same as before. It results, therefore, since the theorems or propositions are deduced from the axioms, that if we take any proved proposition and interchange in it the words "point" and "plane" the resulting statement will be true. Not only this, but we can find other entities, not points or planes in the ordinary sense at all, which satisfy the axioms as if they were points and planes, and to which, therefore, all the theorems apply just as well as to points and planes. In the same fashion, in the non-Euclidean geometry which we mentioned, the so-called *lines* may be understood to stand for semi great circles on an ordinary sphere. The names do not seem really to stand for specific things. We seem to make assumptions and prove results about these *names*, merely as names.

Mathematics in this way reduces to a sort of logical scheme in which we may fit things that satisfy the logical axioms, and pull them out again from the logical theorems. Or, possibly, we need not fit anything at all, but just keep the logical shell, with whatever names may be convenient, and develop it by successive deductions to any extent that we please. Roughly speaking, the axioms become merely definitions of new logical terms, and we can also expand these to any extent that we please. Mathematics becomes identical with pure form—a game of combinations, without regard to content.

Form enters largely into our appreciation of art also. When we study a classical symphony we learn that there is a sonata or a rondo form, or some general scheme in which the music is laid; but, of even more importance, there is a beautiful and complex arrangement of theme, development and counter theme, with a logical order approaching that of mathematics. Without such an order, a long piece of music would be monotonous and trivial. Moreover, the more we study the piece the more relations of this kind we perceive, so that the order forms a sort of minutely ramifying skeleton on which the music grows; and a large part of our estimation of the composer is our appreciation of his skill in this regard.

Having emphasized the relation of mathematics to pure form, we should with consistency be able to look at things with a point of view reversed, and say that wherever we have pure form, we have mathematics. This statement certainly contains some truth. The mathematics need not be carried to such a farcical simplicity as it is in that fad of artificial and obvious painting which is founded on the square root of two, and in which all lines are purposely either the sides or the diagonals of squares; for as we know from what we see in mathematics and music, form may be anything but simple. But from this point of view we must regard seriously an attempt to evaluate our appreciation of form. If pure form is part of the beauty of an object there must in fact be criteria which differentiate pleasing form from the rest. If a curve is beautiful, there must be some abstract element of it which makes its beauty, even though the artist draws it on the basis of his own taste, without analysis.

A recent attempt to investigate this side of the appreciation of art is a book by the mathematician, George D. Birkhoff, with the title *Aesthetic Measure*.[1] We should not cast it aside with a superior smile. It is not consistent to make the hypothesis, as most artists and architects do, that an essential feature of art is in the relations of form, which are themselves pleasing, and then laugh at the practical pursuit of general and theoretical criteria by means of which form may be judged. If the "golden section" is good, is it not permissible to speculate on why it is good?

The author is not proposing to make works of art by a formula. He admits, with us, that analysis comes after the fact. This is a statement which may give us pause when we come to make some additional remarks about mathematics. Nevertheless, unless we admit, as of course some do, that all critics of art and literature lead an aimless and wasted existence, we must hope that analysis of art is possible and instructive, and that there is justification for the attempt to analyze form apart from the object in which it is embodied, and give it its aesthetic ranking.

The figure by which we illustrate our discussion is a simple combination of squares and rectangles, equal segments of lines, and right angles arranged with symmetry.[2] When we sketch in a few arcs of curves with reference to these lines, we obtain a Ming jar of repute (Ming dynasty 1368-1644). The curves themselves, without the rectilinear framework, suggest these many relations of di-

[1] Harvard University Press, 1933.

[2] *Op. cit.*, pp. 80, 82; there the careful drawings are given from which the above sketch is reproduced.

rection and dimensions, and the effort of attention, which is necessary in the apprehension of them, is not too great.

Birkhoff's aesthetic measure for a form is in fact high according as the number of simple relations of form is large in comparison to the complexity of the object necessary to produce these relations. He writes a formula: $M = O/C$, to express this comparison, M being the so-called aesthetic measure. In the numerator, the O is evaluated by counting the various types of relations of order and the number of times they occur, weighted according to their importance. In the denominator, the complexity C, for the case of the vase, is regarded as given satisfactorily by the number of points (10) which it is necessary to fix with the eye in order to apprehend the desirable relations. The larger the ratio, the better the form.

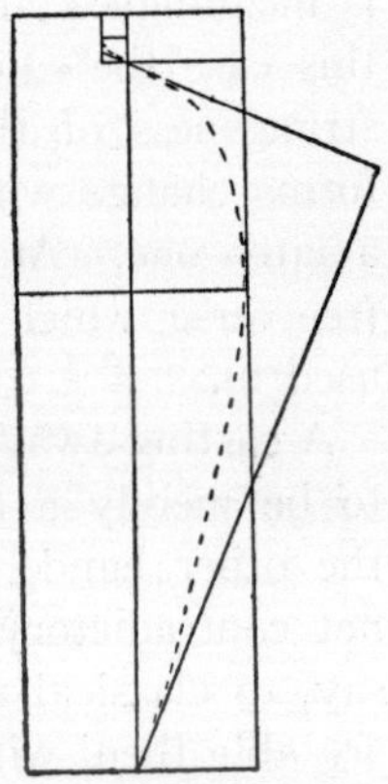

FIG. 1

The vase rendered here is closed in itself; it is a jar or container. Other forms of vases, say, with flaring tops, we may perhaps regard as incomplete, and to require a complementary form of masses in order to complete their harmony. Such a vase provides aesthetic employment for the housewife who owns it. She must "say it with flowers."

In a somewhat similar fashion may be analyzed the structure of a melody or of combinations of chords. Above the line we have the favorable elements of contrast and periodicity, the order of the secondary melodies, and so forth; and below, the complications necessary to produce these relations and the strain necessary to apprehend them. The strain detracts from the pleasure obtained by the perception of the relations, and the aesthetic measure is an indication as to whether the strain, or the pleasure in the relations, predominates.

A textbook on harmony is a discussion of form. It tells us what chords are permissible in what combinations; and we learn these forms by practice, and thus acquire an elementary knowledge of the subject. So perhaps there does not seem to be much point in doing the same thing by a formula. But the author's aim is more ambitious. He wishes to find what the reason is that makes a teacher of harmony permit this and not that, and he thinks he finds not only a criterion for such things, but even one which runs through all the forms of all the various kinds of art. It is a theory of the aesthetic nature of form, as form, without regard to its special content.

An objection to all this theory comes at once to mind, of greater or less weight according to the individual opinion. Some believe that the form of art is, more than anything, tradition. We do not appreciate Chinese music; and likewise, to the Chinese our music is meaningless and disordered. Even though it is possible that in this case the same elements of form are there, but hidden in the strangeness of the sounds, it must be admitted that appropriate forms change with time, and what is new in one age is classic in a later one. At present, for instance, we meet various forms of free verse, which seem to be at variance with structure that we are used to.

A casting away of this formal portion of our art does indeed seem to be merely a deliberate sacrifice of one kind of pleasure. On the other hand, the development of form to new complexities is not contradictory to the general theory. For as we become used, say, to classical music, its complexities become less complex. We are able then, with the same mental effort as before, to apprehend a more complicated structure and we can thus increase our aesthetic pleasure by enriching the form still further. That is to say, a piece of modern music which is so complex as to be barbarous to an untrained ear is not complex to one that has had practice in apprehending simpler forms. Of course, there are limits! Tradition, then, enables us to appreciate new relations and gradual changes by getting us used to the old; and we should therefore expect a gradual development of complexity in a given direction.

Let me present the situation in terms of an oversimplification. We shall represent the complexity of an object by the number of elements, and the relations between them by the number of combinations of the elements, say just two at a time. In this case the aesthetic measure would be given by the formula

$$M = \frac{\text{number of combinations two at a time}}{\text{number of elements}}.$$

Thus if the number of elements is 2, there is one combination and $M = 1/2$; if the number of elements is 3, there are $\frac{3 \times 2}{2} = 3$ combinations and $M = 1$; if the number of elements is 4 there are

$\frac{4 \times 3}{2} = 6$ combinations, and $M = 3/2$. In general

$$M = 1/2 \text{ (number of elements less 1)}$$

and therefore increases with the number of elements.

Roughly speaking, the number of relations of form obtainable increases more rapidly than the complexity of the object. This is what happens perhaps, up to a certain point, with modern music. After that point, the attention required on account of the complexity becomes too great for any but a professional ear, and the relations can no longer be appreciated. Development ceases to be possible, and the relations of form must be sought in a new direction.

The formal technique in mathematics is in the logical structure. Here, too, although many will not believe it, there is the influence of tradition. For if we regard mathematics as a development of logic by the addition of arbitrary definitions and postulates, there is nothing to prevent us from changing the laws of logic with which we are to proceed. That has always been done to some extent. What was good logical procedure to Euler is not always good procedure to us, and even Newton has "proved" some theorems which are not true. Lately there has been much more experimenting in this direction, and the "free logic" of the present day is a fearful and a wonderful thing!

For some mathematicians or logicians mathematics has thus been reduced into a mere game with marks on paper. Each makes his own rules and plays the game his own way. But it is hard for us to believe that after so many fruitful centuries the subject is to degenerate into a hocus pocus. I should say myself that pure form is as fruitless in pure mathematics as it is elsewhere.

I think we can find the key to our difficulties by looking again at the nature of appreciation. For it is not form alone which makes a work of art appealing, although form is desirable and even essential; it is even more in the associations and meanings connoted by the work of art. Let me quote the words of Professor Birkhoff:

> Music owes much of its aesthetic importance to its peculiar emotional effect. This attribute is readily understood. The human voice is a primary means of human expression, and at the same time is a musical instrument. All of us become accustomed to its musical tones, and simultaneously learn to appreciate that which the voice expresses. Thus musical tones have been intimately associated with emotional

feeling. [According to our general theory] music will be most affective when it unites surpassing beauty of form with effective suggestion of emotional utterance.[3]

For example, it is not poetically suggestive, however accurate, to say that from the Berkeley hills at night the lights of Sausalito look like a bundle of fish eggs!

In a Wagner opera such as *Die Walküre* there is an exceedingly complicated form based on themes which are used as symbols, themes for love, for the sword, for Loki, for Hunding, and so on. These themes are almost immediately grasped by the listener and associated with their objects. But the music is no mere procession of titles, like an index or a table of contents; nor is it mere description, like a tone poem with a subtitle such as "kittens playing by a brook"; it is organically welded into a structural whole, its suggestion is more and more rich, and the final scenes build up a climax tragic in its emotional weight. We are conscious of the fire theme; but the symbolism is merged as Wotan's voice dies away, in the significance of the music as music.

There is here an appreciation of the content of the music as well as of its form. But even in this most pictorial kind of music, we lose most of what the composer intends if we hear nothing but fire licking the rocks. The content is not to be translated into words or pictures. It may or may not be strongly emotional. The composer has some elementary musical idea, and he develops it in all possible ways that have meaning to him, by making variations, by setting one part against another, by developing the relations of the parts, by all sorts of manipulations of the form, thus trying as far as possible to come to a realization of the complete content of his idea. It is this aspect that interests us most profoundly.

In mathematics the activity is much the same. An idea which is not clear has the germ of clarity in it. We work it out, guessing where we can and proving where we can, until it no longer has contradictions or surprises for us, and all its parts are knit together. The definitions in mathematics are often relatively simple, and, as we know, everything should be a consequence of the definitions. Yet it is not sufficient to state definitions. It is the uttermost in triviality and banality to say that in mathematics, since one's re-

[3] *Op. cit.*, p. 87.

sults are merely logical consequences of the postulates, one gets out only what one puts in. Exactly the same triviality applies to art in any form. One gets out only what one puts in. It all depends upon what one puts in. After concepts have been developed so that they seem to satisfy, and rest, perhaps stagnant for generations, some new mind sees them confronted with other concepts and in a new light; and a new theory bursts forth. It is as if facts were generated by the rubbing together of concepts.

We are familiar with the story of the ancient painter whose king gave him a prize because he had painted a curtain so accurately that the king had tried to raise it, or of another painter who was chagrined because the birds had tried to peck the painted grapes held by a painted boy—if the boy forsooth had been well painted the birds would have feared to come near!

Or again, at a country fair, a buffoon imitated the cries of animals. He ended by squealing so like a pig that he delighted his audience. But one farmer came by who wagered that he, the farmer, could do much better, if he had just one day of practice. So the next day when they were at the fair again the countryman appeared on the stage, and, putting his head down, squealed so hideously that the spectators hissed and threw stones at him to make him stop. "You fools," he cried, "see what you have been hissing," and held up a little pig whose ear he had been pinching to make him utter squeals. "Men often applaud an imitation and hiss the real thing," says Aesop.

But the audience at the fair had justification for hissing. The real pig's squeal had no place in the holiday booth as an aesthetic creation. Art is a synthesis.

What, then, is its content? Similarly we ask, what is the content of mathematics? It does not matter to me whether mathematics is art or not. Apart from what may be in a name, it is undeniable that music and poetry yield emotions which have no place in mathematics, and perhaps painting also, to a lesser degree. It would probably be farfetched to maintain that a superb painting and a page of Poincaré have the same effect on one who is capable of appreciating both. Nevertheless, the answer to either of the above questions about content affords a suggestion for the answer to the other.

Both artist and mathematician are pursuing the object of making clear an initial concept which has to be worked out. The working out is guided by an intuition of the completed object, by intuition being meant (quite untechnically) a rough foreknowledge of the result, a more or less accurate or hazy mental image. Gradually, and (if the concept is not trivial and the form is restrictive) with intense effort, the image becomes more definite, and the essential details fall into place. The question we are asking now is, what is the nature of the object, which, in order for the work to be appreciated by others than its creator, has to be significant?

In the mathematics of early antiquity the first objects were definitely of the physical world—astronomy or geometry or mensuration. This does not mean that ancient mathematics is limited to trigonometry. For the pursuit of the elementary properties in the measurement of right triangles leads to the fact that the diagonal of a square cannot be measurable rationally in terms of a side: that is, that there is no length whatever, no matter how small, which can be chosen so that it will go just a whole number of times into both side and diagonal. This fact was the secret of Pythagoras. The whole nature of number thus becomes important—what are the numbers with which we count, and what are the numbers by which we measure? You see in retrospect how from the time of the early Greeks such concepts pressed their demands for clarity on the minds of the eager barbarians. The theory of numbers, algebra (in which the language was long entirely geometrical), and geometry were all objective, from this point of view.

Newton attacked seriously the problem of the behavior of continuous media, hydraulics, elasticity, flow of heat, and so forth, and for the investigation of such ideas introduced the study of differential equations. This subject, in the opinion of many, has been the central theme of mathematics for the last two centuries, and its developments and variations still leave us at the present day a host of tempting and difficult problems. But as I suggested at the opening of this address, these are not isolated problems. We look at them with all the experience garnered from the time of the Greeks until the present day. They have a richness which is inspiring.

Mathematics takes its departure from concrete situations, and is continuously refreshed by them. But it draws back from a too slavish imitation of the physical world; it creates combinations of ideal objects, by successive stages of abstraction. The natural sciences lie upon the bosom of the earth, the more theoretical sciences hover closely above them; it is only mathematics which is free and creative. It still seeks the truth, but it is the truth about the truth about the truth . . . of objects.

Let me mention at this point an easily comprehended theorem which exhibits the character of withdrawal from slavish imitation, and, at the same time, formal elegance. It is known as "the geometric theorem of Poincaré."

In the diagram below we imagine a displacement of the points of the plane ring T, among themselves, in such a way that points on each boundary remain on that boundary, points close together remain close together, and no part of the ring is covered twice or fails to be reoccupied by points. In other words, the displacement is a continuous one-to-one transformation of the points of the ring into themselves, leaving the positions of the boundaries unchanged. If on one boundary the displacement is clockwise and on the other counter-clockwise, and if the area of a small piece of the ring is unchanged in value by the transformation, then, according to the theorem, there must be at least two points of T which are not moved.

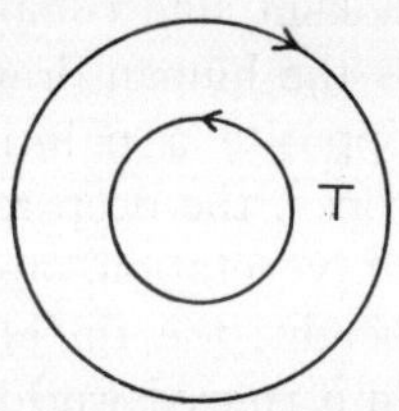

FIG. 2

There are relations of order in the theorem itself apart from the proof, and the complexity is indicated by "the tension" necessary to understand the theorem; that is, to comprehend the hypotheses assumed in the figure.

The theorem came originally from the consideration of motions of a dynamical system which may repeat themselves, as in the particular case of the motion of the sun, a planet, and a small satellite. The statement that there are unchanged points in the geometric situation was seen by Poincaré to be equivalent to the statement that there must be corresponding periodic motions in the case of this celebrated problem of celestial mechanics. The theorem, as it happens, was not proved until after the death of Poincaré, and the proof was ultimately given by this same Birkhoff.

We notice the birth of the problem in contact with nature, the clarification of the concept which is thus withdrawn from mechanics into geometry, so that it is not any longer an attempt to describe a particular event in the physical world, and the resulting generality of the ideas which enables the theorem to be applied on the one hand to geometry and on the other to celestial bodies and billiard balls. The theorem is classic in form and fine in the logic of its proof. Its content and effective suggestion are so vast as to be amazing. Of what importance is it, from the point of view of mathematics, that it can be reduced logically to definition and syllogism? It surely would never have been discovered merely from a knowledge of logic. From every point of view, except the formal retrospective one, it is more than mere logic!

I maintain that this is also the characteristic of the artistic endeavor. It seeks not a slavish imitation of objects, and, indeed, we scoff at such a shallow interpretation of man's relation to the world. Rather, also, it refines itself in successive stages of regression and withdrawal, renewed and revitalized by occasional fresh contact with the world, but essentially free and creative, seeking, by construction, the truth about the truth about the truth . . . about the world and man's relation to it. Conscious of its symbolism, and compressed by its form, unconscious of its purpose, it is the human drama in miniature. As I write, I study on the wall opposite a painting of the Big Thompson Canyon—the colorful rocks, the deep ravine, rhythmically composed with sparse patches of vegetation, and, behind a crag at the very topmost furthest height, the tip of an evergreen against the sky—man's aspiration in a rugged world.

It is quite likely that the painter did not have that idea in mind. But he is trying with all the force of his personality to tell the truth about something, without much thought for his gallery—as though he were imbued with the spirit of the famous argument that if an idea is perfect the existence and immortality of its object are consequent attributes of that perfection. As Thornton Wilder relates, "Uncle Pio and Camila Perichole were tormenting themselves in an effort to establish in Peru the standards of the theaters of some Heaven whither Calderon had preceded them. The public for which masterpieces are intended is not on this earth."

Creation is novelty and novelty seems to be an essential character in the world. For some persons, science and mathematics seem to stand as the antitheses of art, and, by making certain prediction possible, to intend the eradication of novelty. As it used to be stated a generation ago, "if the velocities and positions of all particles that make up the world were given at a given instant, their motions and positions would then be determined at all subsequent times by the immutable laws of mechanics." Perhaps a different formulation would be advanced nowadays, but the intent would be the same. The question is always asked, "How is freedom of the will, how is creative art, possible in a world for which the scientific hypothesis is valid that future events are predictable?" For apparently it is merely a question of time until such prediction will involve mental processes and the entire actions of individuals.

The history of science, however, yields for us quite a different result. In the first place science is wayward; it aims at one goal and instead of getting there goes off at a tangent and reaches another; or, in better words perhaps, when it reaches a goal it finds something different from what it expected; something is left unanswered that is as important as what has been achieved; what answer there may be opens a universe of discourse that is larger than the one in which the question was asked.

We are familiar with the fact that there was an atomic theory in the time of Democritus. But such a theory is entirely different from the present one, for the questions that are asked of atomic theory now were not only undreamed of in the time of Democritus, but they would have been entirely meaningless. When the elementary theory of gases was replaced by a kinetic theory of molecules, it was at first sufficient to consider the gas as equivalent to a number of small elastic spheres. In fact it was not necessary to give the spheres any size at all, as long as they could have mass. The pressure of the gas was the momentum of the spheres, and the temperature was their kinetic energy.

The natural questions as to what would happen then in concrete cases, say, if the gas were greatly compressed, led to experiments which showed quickly that a more refined mechanical model would be necessary. Up to the present time no model has been sufficient. In each case additional experiments are suggested, and sooner or later they have led to unexpected results. Smashing the nuclei of atoms by shooting them with very small high-speed particles is not one of the concepts of Democritus or Leibniz. What would Ernst Häckel think of the principle of uncertainty?

Of course it may be merely an accident that novelty creeps continually into the physical universe, and a day may come when physical theory may not need further radical adjustment. On the other hand it may be that this necessity is a sort of dual image of another sort of wilfulness—that our objectification of the world remains and must remain only partial. At any stage of its development science has only a finite number of concepts at its disposal—in fact, only a finite number of words. It looks, in the light of history, as though a finite number of concepts would not be sufficient.

Mathematics has long been fascinated by the various types of infinity. Of these there are two relatively simple ones, characterized respectively by the collection of all the whole numbers and by the collection of all the real numbers. Or, if we describe them geometrically, one type is like the discrete row of dots, illustrated in the figure below (where the interval on the right of a dot is half the interval on the left), and the other is like the continuous row of points that make up a whole line. In the second type the points are so thickly collected that there is no next one after a given one, there being a point wherever we break the line. But in the first type, given any dot there is a next one beyond, like a pathway of stepping stones.

FIG. 3

It is this kind of infinity that seems to mark the path of science. At any stage there is a next step, but not necessarily a final one; on the basis of given concepts there are always new ones which are necessary; and no gap may perhaps ever be filled completely. The situation is always just beyond finiteness and complete description, and holds out always the temptation to proceed further.

Art and mathematics share this insidious character. When a memoir of mathematics is done it may seem to be nothing but form, and to reduce (as far as the logician is concerned) to a logical skeleton of definition and deduction; but this "seeming" is in the precise sense that a painting may be reduced to a mere pattern of colors. The doing of art and mathematics, and their appreciation, are both creation and discovery. In that way they are practical, for they transform the universe.